The publisher gratefully acknowledges the generous support of Robert J. Nelson and Monica C. Heredia, Marcy and Jeffrey Krinsk, Judith and Kim Maxwell, and Barclay and Sharon Simpson as members of the Literati of the University of California Press Foundation.

Terroir and Other Myths
of Winegrowing

Terroir and Other Myths of Winegrowing

Mark A. Matthews

UNIVERSITY OF CALIFORNIA PRESS

University of California Press, one of the most
distinguished university presses in the United States,
enriches lives around the world by advancing scholarship
in the humanities, social sciences, and natural sciences. Its
activities are supported by the UC Press Foundation and
by philanthropic contributions from individuals and
institutions. For more information, visit www.ucpress.edu.

University of California Press
Oakland, California

Library of Congress Cataloging-in-Publication Data

Names: Matthews, Mark Allen, author.
Title: Terroir and other myths of winegrowing / Mark
A. Matthews.
Description: Oakland, California : University of
California Press, [2015] |
    "2015 | Includes bibliographical references
and index.
    Identifiers: LCCN 2015041749 |
    ISBN 9780520276956 (cloth : alk. paper)
Subjects: LCSH: Terroir. | Viticulture.
Classification: LCC SB387.7 .m38 2015 |
DDC 634.8—dc23
LC record available at http://lccn.loc.gov/2015041749

24   23   22   21   20   19   18   17   16   15
10   9   8   7   6   5   4   3   2   1

So long as authority inspires awe, confusion and absurdity enhance conservative tendencies in society. Firstly, because clear logical thinking leads to an accumulation of knowledge (of which the progress of the natural sciences provides the best example) and the advance of knowledge sooner or later undermines the traditional order. Confused thinking on the other hand leads nowhere in particular and can be indulged indefinitely without producing any impact upon the world.

—Stanislav Andreski

Contents

Preface

In addition to the pleasures available in the consumption of its fermented juice, the grape has long been made special in art, literature, and of course, commerce. The stories woven about both the wines we drink and the grapes they come from can themselves become intoxicating. This book focuses on the latter, the popular and often passionately held explanations for the vineyard origins of fine wines. The following pages present a review of vineyard concepts that flies in the face of most writing on the nature of grapes, but resides well within the boundaries of conventional plant biology.

My personal background may have contributed to an ability to hold the myths regarding the nature of fine winegrapes at arm's length. I am not of a wine-producing tradition. My first encounters with vineyards were as a teenager working table grape harvests northwest of Phoenix, Arizona. I gained an introduction to fine wines during college working as a bartender in supper clubs in Phoenix and Tucson, and in college and graduate school I found my path in environmental crop physiology and water relations. I came to the University of California, Davis and the world of winegrowing well trained as an agronomist and plant physiologist, but as a novice when it came to grapes and wine.

As I gained experience in the world of viticulture, I found that some of the received archetypes were incongruous with elementary crop science. For example, there is a long-standing argument that one cannot both irrigate vines and produce fine wines (yet rain and irrigation water are the same to grapevines). As I encountered more beliefs regarding how to grow fine wine, I grew curious about what was truly known about the grapevine, and how it had come to be known. In the pages that follow, I use direct quotes when introducing concepts, not to single out individuals, but as examples to demonstrate that my arguments were not generated against "straw man" myths.

For much of the received wisdom on the nature of winegrapes, there is a disparity between the passion behind each belief and the supporting evidence. These discrepancies motivated me to conduct a variety of experiments in the vineyard, first on the role of berry size in fruit quality, then to test the high yield–low quality paradigm, investigate the basis of terroir, and finally to assemble my thoughts into this book.

On this journey, I have lost relationships with some colleagues, first when it became clear that I doubted the concepts of terroir and vine balance, and the effect was magnified when I wrote in question of the belief that one must minimize yield in order to produce fine wine. One day, shortly after publishing a paper with data at odds with the dogma that high yield causes low quality, I found a note in my university mailbox from an administrator. It said that a reporter was looking for me, and that this reporter "seemed to think that [I was] saying things that [I] probably [didn't] want to say."

It's important that this discussion about the basis for grape-growing paradigms takes place. Fine wine is inextricably tied up in culture and tradition. We bring our cultures and backgrounds into wine tasting, and we each have our own tasting experience; however, those need not (and indeed *should not*) drive our understanding of the grapevine. The models of wine quality are cultural and ephemeral, but plant interactions with the environment are not. Still, progress in the science of plants and grapevines has gone largely ignored in the tradition-bound popular world of wine and wine marketing. Wine and what we know about wine has changed, while the stories repeated in coffee-table books to describe how and why certain wines are deemed the best remain unaffected. It is time to critically and empirically evaluate terroir, yield, and other putative keys to producing the best winegrapes.

The casual observation is an unreliable assessment that falls prey to what *Uncommon Sense* author Alan Cromer calls "egocentric thinking" and psychologist Daniel Kahneman describes as "thinking fast"— that is, assuming we are right because we are experts. Conversely, it is an objective of science to be rid of bias, and as such, this book is a systematic review of the factors that affect the grapevine that produces the grapes that become the wine.

As far as we know, the grapevine has no sense of political boundaries, tribal skirmishes, or wine style conventions. The grapevine can, however, be bumped and experimented with in ways that provide empirical evidence about how it interacts with the environment. Varying light, temperature, water, yield, and so on and carefully recording the responses in grapes and wines provides data that are important in their own right, but have added importance because the grapevine in the vineyard is noisy (variable) and difficult to assess without the aid of measurements. Measurements either confirm or refute our casual observations of correlations of our tasting experience. By grounding our sense of how the grapevine operates in objective truths, we can develop what plant geneticist Barbara McClintock called a "feeling for the organism," a more intimate and knowledgeable relationship with the grape.

Control of the models of wine quality is being contested by producers, regional interests, and quasi-government institutions. There is often significant hazard in talking apparent science when actually doing politics. Adopting an unsubstantiated basis of wine quality puts the credibility of the wine industry at risk, particularly as today's wine consumer is increasingly analytical. This book is my best effort to reconstruct how terroir and other widely held concepts of vineyard management have played out in the written record, and to hold those ideas up to the available objective and empirical evidence. My hope for the book is that growers, winemakers, and wine consumers will be freed from misguided constraints on where, how, and how much to produce. Moving forward, enhanced study of vine-environment interactions could give rise to new models of wine quality.

There are many people to thank.

For my cherished introduction to the grapevine: Mike Anderson and Hans Schultz.

For their hard work in sorting out aspects of how grapevines function in making winegrapes: my former graduate students—especially Dawn Chapman and Gaspar Roby, whose work was aimed directly at the issues of yield and berry size, and Will Drayton, who contributed to early investigations into the Ravaz Index.

For early discussions about what is knowable about grapes and wines: Bob Adams, Matt Courtney, Tyler Thomas, and Matt Villard; and for creative help with diagrams, Greg Gambetta.

For essential help with French sources and translations: Paul Anamosa, Marc Benassis, Axel Borg, Will Drayton, "Lucie" Fontaine, Abe Jones, Sophie Mirassou, Philippe Pessareau, and Joe Wehrheim, and Elsa Heylen.

For feedback on drafts of parts of this book: Tony Cavalieri, Merilark Padgett, and Elisabeth Sherwin.

For proficient and persistent help with the figures and references: Jiong Fei.

For ongoing support: Barbara Sherwood, my friends on the Sound, and the guys at Saturday noon.

And finally and most important, this project would never have been completed without the enduring, patient, and meticulous assistance of Jessica Makolin with all aspects.

Introduction

Wine is a traditional and cultural product, and most of what viticulturists and winemakers do is attributable to historical and traditional causes. Practices such as pruning to short canes, so that the next vintage's crop is carried in an accessible position, have served the winegrower well for millennia. However, mixed up with commonsense practices are uninformed beliefs, as well as both accurate and mistaken explanations for what makes good grapes and wine. This is inevitable. As humans, we are inclined to explain what we experience and "the nature of things," whether we have sufficient information to generate an accurate explanation or not.

Winegrowing is a convenient term for the growing of winegrapes and finishing to wine, and there is good evidence that winegrowing dates back almost as far as the beginnings of civilization itself, perhaps to 7,500 years ago.[1] Regardless of the specific era of their emergence, it is clear that winegrowing activities have been with us for a very long time. In contrast to winegrowing, the institution of science that tests the basis of ideas in an objective reality originated with the Scientific Revolution about 400 years ago. That leaves thousands of years for explanations of the origins of fine grapes and wine to accumulate prior to the onset of science. Of course the coming of science did little to prevent the development of further mythical beliefs. Some of those beliefs come to us today as received knowledge in winegrowing, explanations for what we do or how the grapevine works that have been handed down through generations. The academics and students in the viticulture and enology programs around the world seek out evidence for or against the received knowledge in winegrowing.

As I prepared to teach the principles of viticulture at the University of California, Davis (UC Davis) several decades ago, it was difficult to find well-documented observations to support some of the common principles of winegrowing. For the first few years, I dutifully taught Winkler's "principles of pruning" (from *General Viticulture,* a recognized authority), but some of the ideas were not only poorly supported by data, a few were also internally contradictory. I had similar experiences with other traditional explanations of how one arrives at fine winegrapes, such as the concept that low yields are required for high-quality fruit and wines. As I scratched the surface, I found some principles suspect, with explanations that just didn't fit with what we already know about how plants work. For example, the most well-known idea within the received knowledge in winegrowing is "terroir"—in which it is often assumed or implied that flavors are transported into the winegrape berry from the soil, but this is unlikely based on the way that the berry is connected to the soil.

Nevertheless, the traditional concepts form a series of principles, myths really, that the world of wine claims should guide winegrowing.[2] The myths live in wine shops, the popular press, and the scripts written for wine tours and tasting rooms. There is a natural tension between the pull of traditional explanations and the push toward new understanding that leads to improved winegrowing. Even in the universities, hallways and classrooms continue to be filled with traditional explanations of how fine winegrapes are produced. For reasons that are not clear, plant scientists have also, in some cases, bought in to these ideas without first applying a healthy skepticism or investigating further. Consequently, the general public accepts these intuitive concepts and ideas as established principles and facts. Many students, clearly and keenly interested in winegrowing and motivated to study wine, arrive at the university "knowing" more than what is truly known about the grapevine.

SOME BASICS OF WINEGRAPE GROWTH AND RIPENING

A few fundamentals of winegrape growth and ripening should be established as a context for evaluating the received wisdom of winegrowing. The grape berry is the result of a two-year process, beginning with the initiation of flower development inside buds on shoots in the summer of the first year (fig. 1). Continued bud development in the first year con-

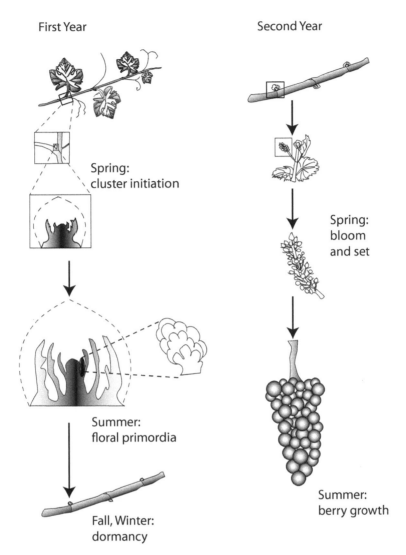

First Year

Second Year

Spring:
cluster initiation

Summer:
floral primordia

Fall, Winter:
dormancy

Spring:
bloom
and set

Summer:
berry growth

FIGURE 1. DIAGRAM OF THE TWO-YEAR PROCESS OF DEVELOPING A BERRY IN TEMPERATE CLIMATES. The physiological decision to initiate a grape cluster is made inside a bud in the summer of the first year, and further development of the bud determines the number of pre-formed flowers. The grapevine goes dormant in the fall. Growth resumes after budbreak in the spring of the second year, and the flowers complete their development and bloom. Many flowers (roughly 50–75 percent) abort; those that are retained grow and ripen into berries. Because two seasons are involved in making a grape, the weather and management practices of two seasons affect the yield.

tributes to more and more flowers on the cluster that will emerge in the second year. Fall brings dormancy, and the process halts until the following spring, when the buds "push" (begin to grow). The shoot grows out into stem, leaves, tendrils, and flower clusters. In late spring or early summer, the flowers bloom, and a fraction of them (25–50 percent) are retained to grow and become berries. Thus, the number of clusters and flowers are determined in the first year, and the final number of berries and berry size are determined in the second year. Yield can be increased or decreased at any step along the way.

The berry of all winegrape varieties exhibits an interesting growth habit in which it grows, then doesn't grow, then grows again (fig. 2A). Ripening begins at approximately the same time as the second growth phase, the transition from Stage II to Stage III. Ripening is a suite of changes that includes softening, increases in sugars and anthocyanins (responsible for color in red grapes), and decreases in organic acids (that give sourness) and in the veggy aroma compound methoxypyrazine (MIBP), as shown in figure 2B.

There are many other flavor and aroma compounds. Tannins, which give bitterness and astringency to wines, are related to anthocyanins as members of a class of compounds commonly referred to as phenolics. Sugars, acids, and tannins are important in all winegrapes, whereas anthocyanins and MIBP are important in some varieties and not present in others. Another class of aroma compounds is terpenes, which are important in some varieties like Riesling and Muscat.[3]

Events like budbreak, bloom, veraison (the onset of ripening), and so on are referred to as phenologic stages or markers. These events happen in sequence as genetically determined, but the specific timing and to what degree the events occur are affected by aspects of the environment (mostly temperature). The aim of phenology is to describe or correlate the timing of specific developmental stages with climatic factors or other timing criteria, often with the goal of drawing comparisons among species or among varieties within a species.[4] Recognition and tracking of vine phenology provides viticulturists with a seasonal framework of what happens when that can be used in scheduling field operations and determining the suitability of certain varieties for specific locations.

Growth and development of the grapevine depend on both the genotype (variety) and the environment (climate, weather, soil) in which the variety is grown. For the grape berries, there is also the additional environmental microclimate (weather conditions around the clusters). These (genotype × environment) interactions alter physiological processes

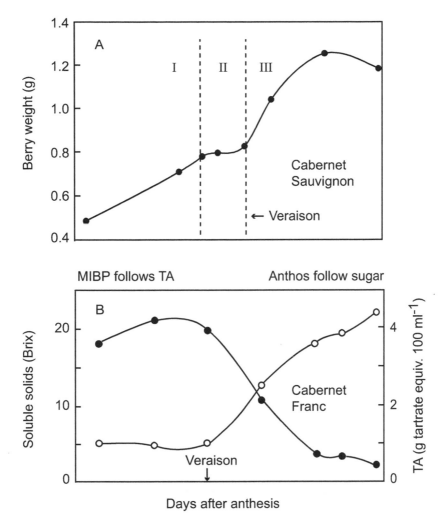

FIGURE 2. THE BASICS OF BERRY GROWTH AND RIPENING. (A) Example berry growth for Cabernet Sauvignon berries. The berry growth habit exhibits a double sigmoidal curve with two periods of rapid expansion. In fruit with this growth pattern, development is commonly described as Stages I, II, and III, with Stage II being the "lag" period between the two growth phases (data from Castellarin et al. 2007). (B) The general developmental pattern of important berry solutes. During ripening, the berry accumulates soluble solids (sugars) and anthocyanins in a roughly similar pattern (open symbols), and loses the organic acid malate and MIBP in a roughly similar pattern (closed symbols). The loss of the malate is reflected in the titratable acidity (TA). The onset of ripening corresponds to the transition from Stage II to Stage III in berry growth. Days after anthesis is equivalent to Days after flower or Days after bloom. (Data from Matthews and Anderson 1988 and Koch et al. 2012 contributed to the figure.)

FIGURE 3. KLEBS' CONCEPT FLOW CHART. Klebs' Concept forms the foundation of our understanding of the traits observed in grapevine varieties and in harvested grapes. In the concept, heredity and environment operate cooperatively through physiology and metabolism to determine the quantity and quality of growth. In winegrowing, the genotype (variety) establishes the potentials for winegrape growth and composition, which are partially modified via the responses of plant physiology to the environment (temperature, light, water, mineral nutrition, etc.). The results of those interactions become the phenotype: the grapevine and its fruit. There is no direct path from the soil (or any aspect of the environment/terroir) to the amount or quality of fruit that bypasses the variety or physiological processes. (Diagram derived from Kozlowski and Pallardy 1997.)

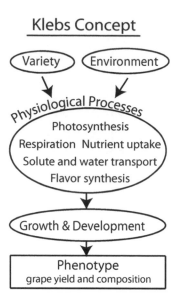

Klebs Concept

Variety | Environment

Physiological Processes
Photosynthesis
Respiration Nutrient uptake
Solute and water transport
Flavor synthesis

Growth & Development

Phenotype
grape yield and composition

such as growth and ripening to produce the phenotype—the traits (such as berry size and color) that we get at harvest. This fundamental truth was elaborated by Georg Klebs (1910) in what is called Klebs' Concept (fig. 3).[5] This is sometimes written in equation form: phenotype = genotype + environment + (genotype × environment), where the environment includes soil and aerial components and within those both biotic (e.g., pests) and abiotic (e.g., temperature) components. Some traits (and their genetic basis) are common to all varieties—these shared characteristics are what makes a grapevine a grapevine. Each variety is distinct because of traits that are hardwired into the fruit, inescapable consequences of the metabolism and physiology driven by the genes that comprise that particular variety. Other traits derive from variety-dependent interactions with the environment.

Klebs' Concept acknowledges the ubiquitous nature (genotype) versus nurture (environment) question. The answer is almost always that both are important, and it is the same for winegrapes. The variety[6]—for example, Pinot noir—is the genotype, and its genetic structure determines what can develop as traits, such as yield, color, flavor, and so on. The grapevine senses many environmental conditions, and those conditions produce signals that regulate the Pinot noir genetic potential via physiological processes into each vintage's grapes.

Quantifying some important traits of the grape phenotype, such as sugar concentration (Brix), color, pH, and TA (both of the latter are measures of acidity), is relatively straightforward. These components occur as dissolved solutes in fruit cells or wines. Sensory attributes are the amounts of simple (sweetness) or complex (green apple flavor) characteristics in fruit or wine that we perceive from the presence of those solutes. The concentrations of the solutes provide reliable information about the grapes and are usually effective predictors of aspects of their corresponding wines. For example, fruit sugar and acid concentrations accurately predict a wine's alcohol and acid concentrations. If such connections were not apparent, then the conclusion that wines result from traits in grapes would be in doubt.

MYTHS AND CRITICAL THINKING IN VITICULTURE

This book takes the most popular explanations for what's in the bottle that came from the grape, evaluates the sources, and probes the literature for empirical evidence in the grapevine and wines. The alleged problems with high yield and big berries are tackled first. Both are thought to be the scourge of fine winegrapes. Roman and medieval legends are used to bolster the idea that winegrowers have long known the benefits of growing a small crop of grapes. There are no legends supporting small berries. Rather, an assumption based on the geometry of spheres is typically the first "evidence" offered. That geometry is thought to lead to flavor and color dilution when the berry is crushed for fermentation. For some, wine quality is all about concentration; both high yield and big berries are thought to lead to dilution of flavor and color.

Next is vine balance, which is also attributed to ancient or a priori origins. Vine balance emerged as a popular guideline to fine winegrapes in the latter part of the twentieth century when academics began to suspect that simply growing low yields was an inadequate approach to good grapes. Viticulturists almost uniformly refer to vine balance, or the ratio of leaves (or pruning weight) to fruit, as the fundamental objective for fine winegrowing. A suite of ripening myths follow, including the existence of a critical ripening period, that ripening is too fast in warm weather, that physiological maturity of the grapes and a critical harvest decision are key to fine winegrapes, and that stressed vines make the best wine.

These adages about how grapes ripen are mostly modern, although based on the historically difficult job of getting grapes ripe before weather and disease ruin the crop. Finally, the biggest myth of all—that "terroir" is the source of fine wine flavor—is again thought to derive from the earliest winegrowing activities and to speak a fundamental truth about the grapevine. It is also probably the most controversial myth of winegrowing.

The common beliefs about winegrowing have arisen and are sustained in the same ways as beliefs about many other endeavors: through intuition, the received knowledge of tradition, the opinion of authority, and as a result of trial and error.

Tradition

Tradition is cultural, comforting, and effective for marketing. "We made this the old-fashioned way" is a claim that resonates with consumers across industries and products. Knowledge of tradition may positively affect one's subjective experience—for example, knowing that a vineyard was managed or wine was made "the old-fashioned, traditional way" may cause the vineyard or wine to be more revered or cherished. But the essence of tradition is repetition. A conservative winegrowing narrative that undergoes (or acknowledges) little change helps produce a "traditional" product. Yet treading grapes with bare feet was a traditional and cherished practice only until it was replaced by something better. Maintaining tradition without thought or investigation precludes any potential progress. With thought and reflection, however, traditional explanations might be affirmed, or a path might be opened for migration toward something better.

Authority

Traditional and authoritarian explanations are related, in that those putting long-standing ideas into practice can take on an authoritative role. We learn about wine quality from authorities in our culture and from our own experiences. Earlier in history, fine wine was what the pope or bishop drank, or what the duke drank. Then the British wine market became highly influential—the preferences of the British helped make certain regions and wine styles revered. We also learned that fine wine was what the famous winemaker or wine writer drinks. According to Émile Peynaud, French enologist and wine expert who has been referred to as the forefather of modern enology, the winemaking establishes the quality,[7] but clearly, the winemakers have a conflict of interest. Perhaps that fact has led to the models of fine wine becoming increasingly influenced by wine critics.[8] Wine writers James Halliday and Hugh Johnson argue that the evaluation of fine wine quality should be left to them, the wine-drinking professionals with vast wine-tasting experience.[9] They have perhaps the greatest familiarity with the greatest number of wines, and enjoy access to a variety of high-end wines.

Wine critics have honed their palates to detect small differences in wines; however, because wine quality involves social and personal opinion, the assessment of quality is quite ephemeral and subjective. Our task is easier when we evaluate the received knowledge about vineyard practices, since we address not the wine per se, but instead investigate the explanations for how it came to be. In very important contrast to the challenges of wine evaluation, the responses of the grape to cultural practices (for example, canopy management practices that alter the ratio of leaves to fruit) are not a matter of subjective experience—the grapevine readily lends itself to quantitative measurements. Changes in fruit composition during ripening, or when grapes are exposed to more or less sunlight, can be objectively known. Therefore, the received knowledge of winegrowing can be held up to empirical evidence—when, of course, it can found.

Trial and Error

Trial and error is a way to make progress empirically—that is, to solve a problem based on experimentation and experience. Among the many myths of winegrowing, a common refrain is that winegrowing regions with a long history and high reputation have refined their production to perfection through hundreds of years of trial and error.

The purpose of trial and error is simply to make something work. It may require long periods of experimentation to arrive at a solution. A potential advantage of trial and error is that it does not require a lot of knowledge; however, trial and error is not a method for finding either the best solution or all possible solutions. The process does not address why a problem was solved, and as such, trial and error cannot provide valid explanations for why a particular solution seems to work. This missing explanatory knowledge is essential for predicting when and where a solution can be applied outside of the specific conditions from which the solution was generated, except of course by luck. Trial and error provides little information that is transferable to other circumstances or environments.

It is difficult to assess the relative contributions of trial and error versus science to winegrowing practices in place today. Both are important, and work goes on via trial and error in vineyards and wineries every season. For decisions based on trial and error and intuition, difficulty arises when one wants to explain why or how practices are effective, and whether they can be extrapolated to other sites and conditions. Any reliable theory or explanation of what makes a good winegrape requires a more systematic investigation of underlying principles in biology.

Contributions from Experimental Work in Vineyards

It is also difficult to establish new facts scientifically. Distinguished irrigation scientist Daniel Hillel summed up his sense of this difficulty in an analogy of tacking in a sailboat against a prevailing wind (fig. 4).[10] One of the several components of the prevailing wind is conventional wisdom, and the prevailing headwind in winegrowing is especially strong. It becomes increasingly difficult to establish new knowledge when "everyone knows" a new idea cannot be true. But, as Descartes said, "The majority opinion is a proof that has no worth for any truths that are at all difficult to discover."[11] We will see that some of the conventional wisdom in winegrowing reduces to intuitive explanations. It may make intuitive sense that low yield and small berries make for better wines, but we need to get in contact with the natural world in order to know about it. Einstein said, "Propositions arrived at purely by logical means are completely empty as regards reality."[12]

On the one hand, there are the easily generated suppositions that come to us as the conventional wisdom of winegrowing. This is represented in figure 4, Hillel's "Path to New Knowledge," by "The Devil"—that is, "the diabolical temptation to . . . venture too deeply into premature speculation and unproved conclusions, in the mistaken belief that Theory alone is Truth and that no more facts are necessary." On the other hand, in studying winegrowing it is also easy to collect masses of data that are noisy and difficult to interpret (represented in fig. 4 by "The Deep Blue Sea"), and this is carried to an extreme in the vineyard, where uncontrolled variables are inescapable.

The search for knowledge that effectively guides winegrowing necessarily involves measurement. Measurements can bring clarity to an issue, but they can also make progress difficult. Every measurement has problems of its own, and often issues arise because definitions are lack-

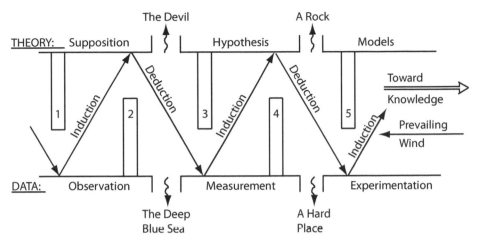

FIGURE 4. IN HILLEL'S PATH TO NEW KNOWLEDGE, TRIALS AND TRIBULATIONS ARE
ENCOUNTERED IN NAVIGATING BETWEEN A ROCK AND A HARD PLACE AND THE DEVIL
AND THE DEEP BLUE SEA. On the bottom is the Data, consisting of Observation,
Measurement, and Experimentation. To some extent this is the cold, hard facts, but it
also consists of lots of data of varying quality. On the top is the realm of Theory,
consisting of Supposition, Hypothesis, and Models, an endless imagining/conceiving of
what could be. The trips from Data to Theory are equivalent to myths—creative
suppositions or predictions of how things might work. They can quickly become
unworkable or untestable theories. The trips from Theory to Data represent deductive
reasoning, whereupon the conclusion necessarily gives rise to a testable specific
prediction. In biology, one can quickly become overwhelmed with too much data to
integrate into a conclusion. The obstacles 2–5 are suggested as various institutional
barriers to scientific progress. (Diagram redrawn from Hillel 1987.)

ing or must be defined more carefully than before. Furthermore, we
recognize a good wine when we taste it; however, we cannot yet know
it in the berry. For example, we have poor knowledge on a chemical
basis of what makes Cabernet Sauvignon grapes become distinctively
Cabernet Sauvignon wine, let alone the subtleties of the environmental
impacts that give rise to fine distinctions among wines. The important
implication is that as viticulturists, we do not know what to measure in
order to predict "wine quality," although we do know some of its components. Resolving the nature of wine back through the vagaries of
winemaking to its origin in the vineyard is a truth that is indeed difficult
to discover.

Wine is fascinating, in part because of its long history, and also
because of the wide range of styles and fine taste experiences that can be
appreciated by those who invest themselves in that experience. Wine

also generates passion, which can intensify enjoyment but at times can also lead to what I call the conceit of intuition—the tendency to assume our thinking is correct (in this case, because we care deeply about the subject).[13] This is another challenge to understanding how fine winegrapes are produced. The received knowledge in winegrowing is a headwind that can inhibit wine and vine enthusiasts from employing the kind of critical thinking skills that are necessary and normally employed for success in other endeavors.

This book explores how popular explanations for winegrowing practices have come to live in the world of wine. Throughout this work, I attempt to track how we have arrived at this conventional wisdom, searching for the origins of terroir and other popular myths of winegrowing. The myths are individually held up to the light of logic, empirical observations, and some contemporary scholarship in viticulture, all in the context of what is known today about how plants grow and develop. Of course, correlation is what we usually have to work with, but a good correlation is not necessarily causation. As long as one can reliably count on one easy observation (yield, for example) to predict another more difficult to resolve phenomenon (fruit and wine quality), vines can be managed accordingly, whether the correlation is causal or not.

There are important distinctions to make regarding wine quality and what we can reasonably attempt to resolve about the myths of winegrowing. First, it is not necessary that we agree on an objective wine quality definition, nor on which wines are best by taste, in order to resolve whether vineyard conditions move wines toward or away from *a definition* of wine quality. For Peynaud, what is objective and subjective about wine is difficult to discern, and aspects of the two overlap.[14] I refer to quality in terms of grape and wine composition, wine sensory profiles, and consensus expert or consumer ratings, each of which contains objective components. The results present a different story than that promulgated in coffee-table wine books and in the rap of wine-tour leaders. In this work, I argue that getting away from these myths when they are not supported by evidence will bring much more craft and creativity to grape growing and winemaking, via improved understanding of the factors that control grape and wine flavor and aromas.

CHAPTER I

Low Yield and Small Berries Determine Wine Quality

One day Mara, the Evil One, was travelling through the villages of India with his attendants. He saw a man doing walking meditation whose face was lit up in wonder. The man had just discovered something on the ground in front of him. Mara's attendant asked what that was and Mara replied, "A piece of truth."

"Doesn't this bother you when someone finds a piece of truth, O Evil One?" his attendant asked. "No," Mara replied. "Right after this, they usually make a belief out of it."

—Zen Buddhist parable

Two of the most widely accepted aspects of received knowledge in wine-growing are that (1) low crop yields and (2) small berries are key factors in producing the best wines. Both winegrowers and the popular wine press frequently invoke the High Yield–Low Quality (HYLQ) and Big Bad Berry (BBB) myths when discussing wine quality in general, or with respect to specific wines. According to highly respected and successful wine producer Paul Draper, of California's Ridge Vineyards, "If there is one common denominator to Ridge's vineyard choices, it's an obsession with old vines, which tend to yield tiny quantities of highly concentrated fruit."[1] David Gates, vice president of vineyard operations for Ridge and a well-respected viticulturist in his own right, reported that everything Ridge does in the vineyard is executed with the goal of obtaining smaller berries.[2] In a *Wine Spectator* magazine cover article where she was pictured as America's greatest winemaker, Helen Turley claimed that low yield is key to wine quality.[3] Wine writers sometimes refer offhandedly to the value of small berries, but Australian wine writer Huon Hooke is more specific in his review of the Irongate Cabernet Sauvignon from the Babitch family in New Zealand: "The quality of the grapes is exceptional. They are small berries with a high skin-to-juice ratio. Consequently, flavor and color are intense."[4]

Some academics apparently concur; for example, according to the website of the Zinfandel Heritage Project, the vines located in the Zinfandel Heritage Vineyard at the Oakville Experimental Vineyard of the University of California, Davis (UC Davis), the vines are trained and pruned as they would have been in the nineteenth century, and "these practices ensure high quality but a low yield."[5] It is more common for academic authors to give a noncommittal nod to the conventional wisdom of low yields or small berries, with comments such as "Crop load adjustment is widely accepted"[6] and "It is often considered that small berries are preferred."[7] However, there are many fully committed references to the BBB notion, including at least one of my own, which will be discussed in the pages that follow.

The popular press claims that the HYLQ wisdom dates back to Roman times; the BBB myth is definitely modern by comparison. Of course yield affects wine supply; thus, it is necessary to consider history and biology (and a little economics) when appraising each of these concepts. Fortunately, growers (and researchers) are able to manipulate both yield and berry size, and the consequences of these manipulations for the concentrations of many berry and wine solutes, as well as for the sensory attributes of the wines, can be measured. A variety of observations from the vineyard are available to assist in these evaluations.

YIELD

Historical Aspects of the High Yield–Low Quality Myth

The line *Bacchus amat colles* from the *Georgics* by the Roman poet Virgil, usually translated as "Bacchus loves the (open) hills," is one early reference point for the HYLQ association.[8] The line has been popular in winegrowing literature dating back to the mid-eighteenth century, and is usually employed in commentary regarding the sites best suited for growing winegrapes. It appears in this context in the two-volume Australian textbook *Viticulture,* in the volume *Practices,* where viticulturist Richard Smart associates the line with a relationship between high yield and low quality.[9] Prominent British wine critic and journalist Jancis Robinson observes that "Virgil may have known that vines love hillsides (*Bacchus amat colles*—because they are less fertile, they produce more concentrated fruit) but many Californians do not."[10] What did Virgil know in Roman times that modern-day Californians do not? According to the "Yield" section of Robinson's *The Oxford Companion to Wine*, "A necessary connection between low yields and high-quality wine has been assumed at least since Roman times when '*Bacchus amat colles*' encapsulated the prevailing belief that low-yielding hillside vineyards produced the best wine. Wine law in many European countries is predicated on the same belief, and the much-imitated appellation controlee laws of France specify maximum permitted yields for each appellation."[11]

In these recent cases, the Bacchus statement is given the same gloss—the hills are preferred because the lower yields result in better wine. Note that it is contemporary authors who supply the reason *why* vines are thought to "love hillsides." The interpretation of Robinson's "necessary connection" with low yield in *The Oxford Companion to Wine* in fact has no direct link to Virgil's words. How did these authors come to know that Virgil was making this connection between low yield and high quality? According to the *Companion*, the line in Virgil reflects a prevailing belief from Roman times that low-yielding hillsides produce the best wines; however, a closer reading of the *Georgics* raises questions about what Virgil may actually have meant by that particular line. Consider the translation of the Bacchus phrase in context:

> Nor do all lands carry all kinds of plants.
> Willows grow by rivers, and alders in dank marshes,
> and the barren manna ash on rocky hills:
> the coast delights in myrtles: lastly Bacchus's vine
> loves open hills, and the yew the cold North wind.[12]

The context of the passage is a discussion of the role of the environment in establishing what grows where. Virgil describes how lands around the world cannot grow all kinds of trees: vines grow on the hillsides, willows by streams, and so on. A little later, Virgil more clearly addresses crops and vineyard cultivation when he says that soils that are dense and tight are best left to corn (i.e., grains), soils loose and light are good for the vine, soils that are rocky can be used for rosemary, and the like for the bees, and that *all three types of soils can be found on hillsides*. Importantly, that passage is followed by this:

> But a rich soil delighting in sweet moisture,
> a level thick with grass, and deeply fertile,
> (such as we're often used to seeing in a hollow valley
> in the hills: the streams flow into it from the high cliffs,
> carrying with them rich mud), one that rises to the south,

and nourishes ferns, hostile to the curved plough,
this will one day provide you the strongest of vines,
and rich flowing wine: from it come fruitful grapes,
and the juice we offer in golden bowls,
while the sleek Tuscan blows his ivory flute at the altars,
and we deliver up the steaming organs in curved dishes.[13]

Now we are in a valley in the hills, deeply fertile and fed by "rich mud" and streams. Virgil refers to the best soils for grapes as giving rise to the "strongest of vines" and "fruitful grapes"—these are not synonyms for low yield. Virgil is evidently describing wines fit for a fine banquet, and promising that much good wine is to be derived from fertile hillside soils. He does mention how difficult it is to plow the hillside, so Virgil may just as likely have been talking pragmatics and longstanding (even in Roman times) agricultural practice: that grains need to be plowed and therefore must be planted on the flats, while trees and vines can be planted on the slopes.

I traced references to the Bacchus line in wine and vine writing back to canonized French chemist Jean-Antoine-Claude Chaptal (of Chaptalization fame) and earlier.[14] The Bacchus quote has been described in winegrowing literature as a Roman proverb, Roman precept, old saying, old saw, maxim, and fact, but in some cases it is not clear that the author was even aware of its origin within Virgil's poem. Often, as in Chaptal's 1801 *Essai sur le vin* (Essay on wine),[15] the Bacchus line is simply reproduced, leaving just *Bacchus amat colles*. When left to context, the sense is a description of a proper place to grow the best wines, with different authors emphasizing soil, aspect of the slope, altitude, or narrowness of a valley.[16] According to my research, it was not until the 1990s that Virgil's comment on hillside viticulture was used to invoke low yield as a causal factor in the resulting wine quality.

Another concern regarding the interpretation of the Bacchus quote as suggesting a fundamental relation between yield and quality is whether Virgil's writing was a well-informed and accurate description of how grapevines operate both then and now. Virgil was one of the most accomplished Roman poets, and as such he wrote poetry at a high level that would certainly have been a difficult read for farmers. Some argue that he wrote in the context of farming but not as a guide for farmers, perhaps instead with the intention of glorifying the rural lifestyle for returning soldiers and/or pining for the good old days in the country.[17] His work on agriculture has at times been cited as a source, but others, even in Roman times, criticized Virgil's poetry for its inaccuracies. For example, Seneca (philosopher and author of *Letters from a Stoic*) wrote: "So says our Virgil, who looked for the most apt thing to say, not the most accurate: he didn't want to teach farmers, but please readers."[18]

Fortunately, several of the best-known Roman writers on agriculture and viticulture also made pertinent comments relative to grape yield and quality. Columella (a friend of Seneca's) was the author of *De re rustica* (Agriculture), Pliny the Elder penned *Naturalis historia* (Natural History), and Cato the Elder, also known as Cato the Censor, wrote *De agri cultura* (On Farming) about two centuries before the others.

Cato the Censor claimed, "Most important in purchasing a farm is whether the vineyard makes good wine and the yield is great."[19] Cato was widely quoted by Pliny, and both Cato and Pliny apparently agreed that a vineyard was the most profitable investment in agriculture, which can still be true today.[20] In an extensive survey of wine regions and grape varieties, Pliny notes that the wine varieties in his second class produce less than the varieties in his top-quality class. Pliny reports on many regions famous for good wine, identifies the best wine as coming from vineyards near the sea, and notes that wine character responds to weather factors, such as whether a prevailing west wind was present during the season. He describes some regions as producing small amounts of low-quality wine and others as producing large volumes of well-respected wines. Pliny does describe one region as being more famous for its productivity than quality; he suggests that the reputation of a top wine region (near Mt. Falernus in southern Italy) is faltering because of an emphasis on quantity rather than quality; and with respect to one variety he says that it makes up in production what it lacks in quality. Pliny cites Virgil often, but not the Bacchus line, nor does Pliny invoke a hillside or low-yield requirement.[21]

Columella distinguishes hills as producing better flavor and lower yield compared to the open spaces, acknowledging that there may be a trade-off of production for quality in selecting some vineyard environments. But Columella's distinction between the wines from open slopes and those from the steeper hills or open spaces is not part of a theme that relates yield and quality or that implies a causal relation. Rather, Columella repeats a story about a farmer, Pavidius Veterenssis, who had "two daughters and a vineyard; when his eldest daughter was married, he gave her a third of his vineyard, and this notwithstanding he used to gather the same quantity of fruit as formerly. When his younger daughter was married, he gave her the half of what remained and still the income of his produce was not diminished."[22]

In discussing soil management, Columella claims that applying manure is the "way to grow not only luxuriant crops of grain but also very fine vineyards." He also believes that "the uppermost roots of vines and olives [would be] detrimental to the yield if they [were] left," and recommends that they instead "be cut off by the ploughshares."[23] Columella also makes specific recommendations for vine training, claiming that those whose chief goal is high wine quality should "encourage the vine to mount to the top of the trees, in such a way that the top of the vine keeps pace with the top of the tree." At the same time, he also indicates that higher supports are needed for higher-yielding situations. Columella apparently feels that good management will improve yields. He argues that low-yielding vineyards are often the result of a lack of investment in caring for the vines, and that low-yielding vineyards should in fact be rooted out.

In contrast to Virgil's low-yielding hillsides, there is straightforward advice in the writings of these Roman agriculturists that promotes intensive, yield-increasing vineyard management practices in conjunction with high-quality wines. This advice is at odds with the contemporary interpretation of the one line about hills in Virgil's poem—which may not be intended to advise on yield at all.

In the end, it is difficult to extract meaningful information about what makes good winegrapes from an anecdote in the form of an ancient poem, especially one full of mythological references. The significance of Virgil's comment for those who repeat it may, it seems, have more to do with its age and its fit in the contemporary wine-world zeitgeist, in which low yield is equated with quality. There are many other passages from the ancients that are not cited in the wine press, on back labels, or in tasting rooms, as they do not fit or benefit today's marketing. Pliny reports that pouring wine on plane (aka sycamore) trees encourages growth, being most beneficial to the roots.[24] According to Columella's uncle, whom he respected, "Dung should not be applied to vines, because it spoiled the flavor of the wine; and he thought a better dressing for making a heavy vintage [high yield] was humus." This same uncle also believed that applying human urine aged for six months "improves the flavor and the bouquet of the wine and the fruit."[25]

Opinions of the ancients on agriculture and plant biology must be received and evaluated with the understanding that at the time, knowledge of plant functions was limited. Even if the best wine in Virgil's time *did* come from hillside vineyards, it was not possible to test or know whether the positive wine attributes were derived from low yield itself, or if they were the result of other hillside factors, such as shallow, well-drained soils. In addition, Pliny describes varieties that produce wines "deemed excellent in their own country, while elsewhere they are held in no esteem at all."[26] What we appreciate in food and drink is very much a cultural phenomenon, and wines in Roman times were also very different from those of today. For the time being, I suggest that we accept that Roman wine was almost certainly bad by today's standards. At the time, oxidized white wines were the most revered by Romans, cooking or heating wines was standard, and although some wines were consumed "neat," most were adulterated with honey, pitch, salt, or herbs in order to make them palatable. Until we find that there is a human universal in wine aesthetics—one that makes the wine judgments of the ancients similar to those of consumers today—the ancients must be considered unreliable as experts on wine quality.

The Disloyal Grape

Another legend that is popular in the wine press involves Philip the Bold and his ducal order to cease Gamay production in favor of Pinot noir in late fourteenth-century Burgundy, ostensibly because Gamay's high yield resulted in low-quality wine. As the story goes, Philip the Bold attributed the (recently) poor wine market in Burgundy to the Gamay grape and its high yield.[27] Some consider this edict a wise predecessor to appellation controls. This legend lives on in popular wine books and on websites around the world of wine as evidence of early insight into fine winegrowing.

Rosalind Kent Berlow, former professor of history at Touro College in New York, looked into the affair in some detail. She reports that a decree issued by the duke on July 31, 1395, called for all Gamay vines to be cut down within thirty days; the duke also forbid the application of organic fertilizers, manures, and other waste, and set a very high fine for either offense. Philip the Bold claimed that the high yield of Gamay was a new problem, and called the Gamay grape "a very bad and very disloyal plant" that was "very harmful to human beings."[28] He also claimed that his edict was being carried out in order to save the depressed wine market, for the good of Burgundy and the welfare of its people. According to Berlow, however,

> The wine market was indeed depressed in the mid1390s, but the causes for this do not seem to have been in any way related to the use of the Gamay or organic fertilizer. In fact, these practices would seem to be logical attempts to

alleviate the problem of declining productivity. The net effect of the ducal prohibitions, however, was to plunge the area into a deeper recession. Productivity declined even more. Speculation in wine sales dropped to an all-time low and the population at large, in both town and countryside, was impoverished.[29]

Berlow suggests that the duke's decree was instead an attempt to curtail the independence of local leaders. It is clear that, despite popular reports today of the despised Gamay, the populace hated the duke's plan to rip out the variety, and were not inclined to follow or enforce it. Eventually, the mayor of Dijon was jailed in order to put the decree into effect. Berlow reported a huge negative effect of the policy:

> The human cost of this policy is stunning. The class of men most directly involved were not the irresponsible charlatans the duke suggested in his ordinance. Rather, they were the political and economic leaders of the community, men who enjoyed considerable local power and respect. The destruction of such families as the Barolets of Saint-Romain and the Bauduyns of Beaune could not have been accidental and one wonders whether or not it might not have been the real purpose of this legislation. With the destruction of the entrepreneurial class came the decline of the wealth of Burgundy. The resources of the Duchy had once been sufficient to propel its dukes to national power. Now, Burgundy would take a back seat in a state which, while using its name, was centered more and more in the alien territories to the north.[30]

Berlow's article, "The 'Disloyal' Grape: The Agrarian Crisis of Late Fourteenth-Century Burgundy," is sometimes cited in the wine press, but usually in a manner that implies support for the HYLQ myth, which is not at all accurate. Despite references to the contrary, her primary conclusion was that the idea of a new high yield–low quality wine problem due to producing Gamay was not supported in her review of court and transactional documents of the time in Dijon.

In *Wine and the Vine: An Historical Geography of Viticulture and the Wine Trade*, Tim Unwin is at odds with the wine press in referring to Berlow's work in a way that correctly represents her thesis.[31] According to Unwin, the situation that developed in late fourteenth-century Burgundy was in part a consequence of the devastating effect of the black plague on available labor. Perhaps Philip the Bold's decree was actually an attempt to push the vineyards into a system that required less labor to harvest grapes and make wine. In 1441, the duke's grandson, Phillip the Good (who sold Joan of Arc to the English for 10,000 gold crowns), is said to have banned winegrowing from specific "unsuitable land"—in particular, the plains or marshland around Dijon. Eminent French historian Roger Dion described the emergence of Pinot noir in Burgundy as the result of a concerted promotional effort of the local dukes, and called the campaign by Philip the Bold to make the reputation of Beaune wines the finest in the world a "propaganda triumph of Burgundy's Valois Dukes."[32] Both Philips, the elder Bold and grandson the Good, are recognized for their efforts to promote Burgundian regional wines. They may have been able to get a better wine price for Pinot noir (perhaps because Pinot wine actually tasted better, but possibly for other reasons such as that it traveled better than Gamay), but the premium was not reflected in the price paid to the growers for grapes. Regardless of these speculations, when scrutinized by Berlow (a scholar disinterested in the romanticized tales of the popular wine press), the popular legend of the disloyal grape came up wanting.

In her article, Berlow showed that Gamay was not a new variety that had come in and made a mess of things. In fact, the variety was present and had been taxed more than thirty years earlier. The use of manure and other organic fertilizers was also not a novelty in 1395. Pliny wrote of the practice as being ancient in his time (AD 23–79), and discussed various manures and their attributes. Other medieval scholars have argued that there wasn't enough manure available at that time to treat vineyards anyway.[33] Thus, it is unlikely that the duke was reacting to a recent change in production (either via variety or added fertilizers) that led to increased yields and low quality.

Berlow's research also shows that production of wine declined from 1371 to 1394 to less than 72 percent of its midcentury level, so production (or yields) had not been increasing in the twenty-three years prior to the edict. In addition to Berlow's arguments against the legend of the disloyal grape based on period records, if the problem with Gamay was that quality was diminished by high yield, then simply requiring lower yields would have done the trick, without imposing the economic hardship involved in ripping out and replanting entire vineyards.

Yield or Supply?

The legends of "Bacchus loves the hills" and the disloyal grape are not reliable support or evidence for early recognition of HYLQ as a truth so clear that it was easily resolved centuries ago. Far better established is the fact that, since Roman times, prosperous periods increased demand for wine, and thus new lands were planted to vineyards, and economic depressions led to chronic overproduction.[34] Unwin and other authors have described conventional supply and demand forces historically active in the wine trade, despite ongoing and fluctuating political controls over production. The concern for adequate demand was apparent in a very early (1602) review of agriculture in France that posed the question, "If you are not in a position to sell your wine, what will you do with a great vineyard?"[35]

Contrary to the HYLQ paradigm, good weather, high yields, and quality wine have historically gone together by many accounts. A comprehensive study of local records of yields, wine quality, and temperatures in western Europe from the fifteenth through the nineteenth century concluded: "The three curves—for harvest date, vine yield, and wine quality—differ in detail from year to year. But overall they show a strong parallelism, rising and falling together with the main shifts in the prevailing temperature regime throughout the summer, from late spring through to early autumn."[36] Parallel increases in quality and yield are reported in shorter periods as well. In what modern French historian Leo A. Loubère calls the Golden Age (beginning in the 1850s), wine merchants convinced the elites to go for the higher-priced releases, regardless of supply. Loubère shows that wine quality and yield had inverse relationships in Burgundy in 1820–50, as predicted by HYLQ, but yield and quality went up together in 1850–79.[37] Poor-quality vintages in Bordeaux in 1927, 1930, and 1951 were attributed to poor weather, rain, and cool temperatures, which inhibited ripening, rather than to high yields.[38] There is no relation between yields in Napa County, California, and published wine scores for the vintages of 1975–99 (discussed below in conjunction with fig. 6).[39] Although limited in number, this cohort of reports is not unusual in suggesting that in the large context of comparing vintages, yield and quality are not mutually opposed in any robust or fundamental way.

Historically, when supply has been relatively high, concern for quality has increased.[40] Wine production controls have often been implemented under the guise of maintaining quality, but it is possible that the controls serve two purposes: to eliminate lower-quality wine *and* to eliminate overproduction, both of which could damage return on investment. This pattern must be considered one of the possible factors in the HYLQ paradigm—the creation of an artificially limited supply, held out as a virtuous sacrifice of yield for the sake of quality. Thus, when cultural historian Maguelonne Toussaint-Samat reports that although the development of brandy production as an alternative destination for winegrapes was helpful in dealing with excess production in the Middle

Price per bottle   Volume of wine produced at a bottle price

FIGURE 5. THE TYPICAL SEGMENTATION OF THE U.S. WINE MARKET BASED ON PRICE PER 750 ML BOTTLE.

Ages, "vineyards sometimes had to be destroyed: a drastic measure, adopted to safeguard quality,"[41] she may have substituted "quality" for "limited supply." (Again, if high yield was the fundamental problem, the growers could have lowered yield and produced fine wine.) Associating high yields with low quality may be confounded in this issue of recurring oversupply, with its companion bad times and the desire (or perhaps economic need) for supply to be limited, in order for the price to be high enough for the winegrowing venture to be profitable.

The supply of wines is generally inversely proportional to wine price point (fig. 5). Of course, price is one of the extrinsic properties taken by many as an indicator of quality, but price does not necessarily reflect quality or yield (which is itself only a component of supply). There are alternatives for the disposal of grapes and wines: distillation, alternate labels, and bulk market sales, and each reduces the supply of a wine from what would have been produced based on crop yield. High-end wines do not follow the normal rules when operating as luxury items or Veblen goods (in which demand *increases* with price). Nonetheless, the supply of a wine may be reduced by restricting yield in individual vineyards by cultural practices such as more severe winter pruning or shoot and cluster thinning during the growing season.

For some winegrowing regions, yield is regulated by law—that is, yield is required to be below a certain limit in order to receive a certain label (geographic indication, or "GI") on the bottle. The traditional story is that the regions were established by soil-based (hence, *terroir-based*) flavors in the wines. How the boundaries for the regions (often called appellations) come about, and what this designation says about the wine and its relationship to the environment of a designated area are topics that will be explored in chapter 4. The appellation concept, a fundamental factor in wine marketing, designates a geographic area for a wine label, serving in effect as a legal description of terroir.[42] The early laws set boundaries and also prevented growers outside the legal area from bringing their wines to market in major trade centers like Bordeaux until after those within the boundaries had the opportunity to sell their wines. When limiting the acreage of an appellation was insufficient to secure a decent price, rules grew to also include crop yield.[43]

The appellation rules were arguably put in place to keep scoundrels making inferior wine from ruining the reputation of the good guys (thus hurting demand), and/or from making more money than those making higher-quality wine. In France, the maximum yield is set in the regulations for each Appellation d'Origine Contrôlée (AOC); and similarly in Italy with the Denominazione di Origine Controllata (DOC) and with Spain's Denominación de Origen (DOC). The maximum yield allowed for a given controlled area such as an AOC or DOC in a given vintage is a base yield of the controlled area—the ongoing nominal limit, as modified by the other limit, in spirit an annual judgment call, referred to as the Plafond Limite de Classement (PLC). Where yields exceed the limit, the resulting wine may be downgraded to a label that implies lower quality, but really only indicates that the winegrowing methods did not conform to the AOC rules.

At first glance, the effect of restricting supply in the presence of demand is to create monopolistic competition (in which the few producers control the supply) and to increase the price. This is a market in disequilibrium, one with excess demand.[44] However, the conventional and simple supply and demand relationship is for products that are similar. Wine producers and the wine press contend that the products at various price points are not similar, and the range of prices that is found within the wine market supports that notion.

When the market beckons, production often runs beyond the nominal limits, even in AOC zones. In most vintages, the PLC allows a production around 20 percent above the base yield, and for years the base yield has been increased.[45] Yields in Bordeaux are reportedly never at the legal limit, and a decade ago it was reported that the trade association in the Côtes du Rhône region hired vineyard inspectors to police growers, to ensure that grape yields didn't exceed the authorized limit of 2.5 tons per acre.[46] The move was part of a crackdown to enforce the regulations governing the AOCs, as producers were finding a market for more wine than was allowed.[47] The area of Champagne production was recently adjusted upward to accommodate demand,[48] and according to the appellation's website, yields have been going up as well.

On the other hand, regulations do limit the yield in the appellation systems, and yields are probably lower than they would be if AOC regulations did not exist. Compared to France, average yields in Australia and Germany are higher. (Although California yields are comparable, they vary over a larger range.) The effects of the HYLQ myth extend to regions within countries, which brings us an important factor of grape and wine prices that confounds a simple HYLQ relation—the environment. In California, for example, yields are higher and prices lower for grapes grown in California's Central Valley than for their higher-priced and lower-yielding counterparts in the coastal valleys. The same phenomenon is experienced, in general, when moving south in France or Italy. The high yield–low price relation is considered by many to be causal, but if wine quality and price were primarily a matter of yield, there would be no differences in yield among winegrowing regions, because yield is so easily regulated.

Wineries of scale usually vie to produce an amount of wine under a label that they expect to sell (i.e., to meet or nearly meet anticipated demand), hence the upward adjustments in yield regulations when demand is strong. In the United States, wine labels carry geographic indications, but there are no quasi-government organizations that set yield limits or other production practices. Most contracts for winegrapes in the United States, however, allow the buyer to determine how much crop to produce, and yield is often reduced by the buyer in the name of quality.

At the vineyard level as well, most winegrape sales in the United States are priced in $/ton, with a set of contingencies.[49] The buyer often controls the yield target, frequently requiring the grower to produce less, and to even remove fruit during the season, under the auspices of quality control. The fruit quality (or the perception thereof) is an important price consideration in horticulture in general, but more so in winegrapes than for other horticulture products. The price per ton for winegrapes depends on the supply, but also on the local/appellation reputation, the track record of the particular vineyard in the eyes of the buyer, and the sense of quality of the fruit from that season (if a price has not already been agreed upon prior to the season). In contrast to a commodity like corn, for which there is a single price at a given time, the current price for fruit from a variety such as Cabernet Sauvignon in California varies more than twentyfold. The yields vary less than the fruit prices, from about 2 to 12 tons/acre, compared to $500 to $12,000/ton. But wine prices vary even more, as much as 50 to 500-fold, and from about $2 to more than $1,000/bottle.[50] With variations like these, something other than yield must be quite important. Famed wine critic Robert Parker claims that those who keep yields low in turn keep quality high,[51] and do so at their own economic disadvantage.[52] From these numbers, however, it appears that one could potentially more than make up for lower yield with higher prices.

It is possible that the reduction in supply caused by the limits on yield is an unintended consequence of well-intended efforts to keep quality high—it is simply thought to be necessary to obtain high-quality wines and to prevent low-quality wines from high-yielding vineyards to damage a region's reputation. Still, it is curious that when demand is low, the "need" for quality control rises. Murky relations between yield, supply, quality, and price persist in the world of wine. Some involved in wine production and wine journalism speak of yield as though it were an antiquality factor. While there may be something to that, a more straightforward truth is that yield is an antirarity factor.

QUALITY AND YIELD ON GLOBAL AND REGIONAL SCALES

What are the consequences of increasing or restricting yield for both fruit and wines, independent of entanglements with wine supply or price? In keeping with HYLQ, wine writers and winemakers bemoan the negative effects of increasing grape yields. James Halliday and Hugh Johnson describe a general problem with wine quality around the world due to supposed increased yields during the latter part of the twentieth century. Robert Parker offered a similar analysis (first in 1997): "For much of the last decade production yields throughout the world have broken records with almost every vintage. The results are wines that increasingly lack character, concentration, and staying power."[53]

That is saying a lot—how did these wine experts come to know cause and effect between vineyard yield and wine quality? First, let's tentatively accept their word as fact on the tasting side and address the vineyard yields. Parker did not cite yield data, but may not have thought that necessary. He went on to say, "The argument that more carefully and competently managed vineyards result in larger crops is nonsense." As an example of the HYLQ concept in action, Halliday and Johnson cite "the German experience with Riesling" to argue that it was increases in yield that caused the well-documented difficulties (damaged reputation) for German Rieslings during the 1980s.[54] They compare yields in the 1970s with those from a century earlier (the 1870s).[55] More objectively, I think an analysis of grape yield over that period of time from any established wine region would show significant increases, yet it was only German Riesling wines that were said to have experienced a damaged reputation (purportedly due to high yield) during that time.

For comparison, I tracked down yield records over the past several decades for Germany and for Cabernet Sauvignon in Napa California. The average yields in Germany bounce around, but with no large increase corresponding to the difficult last two decades of the twentieth century (fig. 6A). Similarly, the average Napa County yield for Cabernet Sauvignon has remained relatively steady during the four decades for which data are available (fig. 6B). In neither case do the data reveal the yield increases that are the supposed cause of supposed losses in wine quality.

Coming back to the quality of the vintages, in contrast to the statements of decreasing wine quality, independent analyses of wine scores from leading wine critics for Napa Cabernet Sauvignon consistently indicate that scores went up, rather than down.[56] And if we plot average vintage scores against the yields of Cabernet Sauvignon in Napa (for 1975–99), the predicted negative relationship between quality and yield is not revealed (fig. 6C). Thus, both the supposed yield increase and its supposed effect on expert ratings are called into question.

To be fair, reliable data at this scale are difficult to come by.[57] The stability of the yields over that period in Napa County is surprising to me, because major vineyard redevelopment took place in the 1990s, increasing the vines per acre, which could be expected to increase yields. However, yield is managed by pruning and thinning, and the Napa County data should be more reliable than back calculations from volumes of wine (which allow for various losses to occur before the calculation). I don't know if there is other evidence that German Riesling yields increased in conjunction with the hard times, but the wine critics agree that there were several top vintages in both the 1980s and the 1990s.

Although this little analysis is neither comprehensive nor definitive, it does present evidence that both the supposed cause (yield) and its

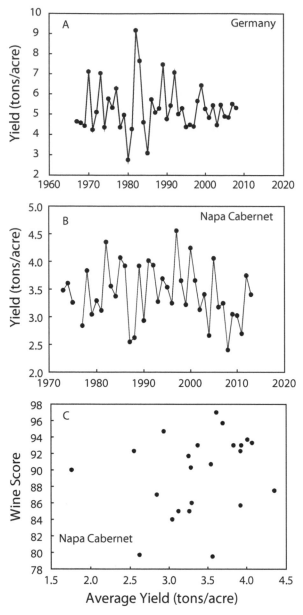

FIGURE 6. TIMELINES OF AVERAGE YIELDS, AND AVERAGE
VINTAGE SCORES. (A) Timeline of average yields of German
vineyards; data from the German Wine Institute. (B) Timeline
of average yields of Cabernet Sauvignon in Napa Valley,
California; data from the Napa County Agriculture
Commissioner (via Michael Anderson). (C) Average vintage
scores of North Coast Cabernet Sauvignon for vintages with
various average yields from 1975 to 1999. Wine scores are
the mean of *Wine Spectator, Wine Enthusiast*, and *Wine
Advocate* vintage ratings for North Coast California Cabernet
Sauvignon.

putative effect (reduced quality) are questionable without more solid evidence. Nevertheless, Halliday and Johnson describe specific negative consequences of high yields in the case of German Riesling: "As yields go up, the tendency is for bunch and berry weight to increase. This in turn leads to higher pH, lower acidity, and lower total flavor content at a given sugar level. In terms of taste, this means lower flavor intensity, less total flavor, and softer wine that will not age so well."[58]

Investigating such objective, measurable parameters as berry size and acidity facilitates more reliable conclusions regarding how the grapevine operates, compared to inferences from the more subjective taste experience, keeping in mind that the overall purpose is to obtain a better understanding of how yield might affect the taste experience. Unfortunately Halliday and Johnson cite no sources, which makes their claims about the grapes difficult to investigate. Halliday and Johnson also do not describe how yield was to have increased, but they imply that it was by grower choice, which further implies changes in pruning to leave more buds, or less cluster thinning to leave more clusters. I found only a couple of studies with pertinent data on berry size in Riesling, none of which supported the claim that size increases with yield.[59] Unless the Riesling yield increased by irrigation or rainfall, the tendency would be for berry size to decrease as the number of fruit increased, as is the general case for fruit.

We can use some work by prominent viticulture researcher Andy Reynolds and coworkers at Brock University in Canada as an introduction to research on the consequences of yield manipulations for fruit and wine. They conducted multiple studies over multiple years to test the effects of yield and canopy management on fruit and wine of Riesling. Although there were differences in fruit composition when yield was changed in various ways (including canopy spacing and cluster thinning), the alterations produced no significant differences among the wines, despite increasing yield up to 100 percent: "Ranges in soluble solids (sugars), TA, pH, and monoterpenes in the juices used in winemaking were quite small, and maximum and minimum values did not correspond to treatment combinations with highest and lowest yields."[60] Reynolds's studies are particularly important, because he imposed treatments in the vineyard over multiple seasons. He also carried the experiment through to finished wines, as well as sensory analysis of those wines by a panel of judges, whose individual consistency and agreement with other judges can be statistically validated. Reynolds's results indicate a low sensitivity of Riesling fruit and wines to large alterations of yield.

What Is the Expected Relationship of Grapes and Wines to Yield?

A visual starting point for examining the HYLQ paradigm is a linear decrease in quality with increasing yield (curve 1 in fig. 7), an illustration of the causal relationship suggested by many proponents of the HYLQ concept. In this straightforward hypothetical, any increase in yield results in a sacrifice in fruit and wine quality (as commonly implied by HYLQ), but this purported relationship is merely a starting point. The relationship between yield and quality could take many forms, including a very steep slope or an almost flat line, the latter indicating little or no dependence of quality on yield.

Some claim to have a more specific knowledge of the shape of this supposed relationship or curve. According to Karen MacNeil, author of *The Wine Bible,* "The relationship of yield of grapes to wine quality is extremely complex and nonlinear."[61] No further explanation of the complexity was offered with that statement, but potentially important insight is given in her introductory comments on yield and quality: "We do know this: For every vineyard, there is a breaking point—a point where too many grapes will cause the vineyard to be out of balance and where the subsequent quality of the wine will plummet."[62]

Halliday and Johnson include a short chapter dedicated to yield and quality in *The Art and Science of Wine,* in which they claim, "However yield is measured, there is no question of relating it directly to quality."[63] Apparently by "directly," they mean something similar to MacNeil, as they cite the "the French rule of thumb" of 3.7 tons/acre (nominally 50 hectoliters of wine per hectare) as a yield not to be crossed when attempting to produce fine wine. Whereas MacNeil sees this threshold as vineyard specific, Halliday and Johnson report this as a broader overall principle indicating, for example, that high-quality German Riesling arises from yields similar to those required for Chardonnay in Burgundy vineyards. It is interesting to note that the Napa average in figure 6B is curiously close to the 3.7 tons/acre standard suggested by Halliday and Johnson. Whether trial and error has worked out a fundamental truth or a herd mentality is in effect is the question.

The hypothetical breaking point, whether for winegrapes in general or for a vineyard, is represented by curve 2 in figure 7, in which the real

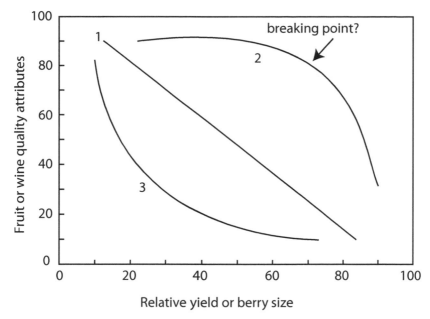

FIGURE 7. HYPOTHETICAL RELATIONSHIPS OF QUALITY AND RELATIVE YIELD OR BERRY SIZE. The conventional wisdom is that quality decreases as yield or berry size increases. The vertical axis can be a single attribute, such as sugar concentration or color in a berry, or a general quality, such as a wine score from a critic or taste panel. Three hypothetical curves are shown; each represents the HYLQ and BBB concepts. Curve 1 represents a direct (linear) dependence of fruit and wine qualities on yield or berry size. Curve 2 represents quality as relatively insensitive to yield or size increases up to a point, beyond which quality falls. Curve 3 represents quality as relatively insensitive to high yield or big berries, but below some yield or berry size quality becomes increasingly positive with further decreases in yield or size.

action in loss of quality occurs above some specific yield point. If the relationship of quality to yield indeed follows curve 2, reducing crop below "60" would have no effect on the wine, and hence sacrificing yield would hardly be the virtue that it is generally held out to be (in contrast to curve 1, where every reduction in yield brings an increase in quality). Additionally, if that hypothetical threshold (or breaking point) of curve 2 were known, producers could exploit this knowledge to produce a higher volume of similarly fine wines, rather than needlessly restricting production.

If, alternatively, the quality response is that shown in curve 3, yields above some intermediate value will all produce similar and relatively low-quality wine, and only at very low yields would each incremental decrease lead to an increasingly better wine. The relationship in curve 3 reflects the extreme importance of low yields for obtaining high quality proclaimed by some wine writers and producers (and printed on the back labels of some California wines). Because the yield must be so low, very little of the high-quality wine could be produced and sold. Curve 3 expresses a very different relationship between yield and quality than that described by MacNeil for vineyards or suggested in a more general sense by Halliday and Johnson.

MacNeil gives a measured evaluation of the role of yield in her glossary, noting several caveats with respect to the HYLQ model. Yet, when discussing the designation of Grand Cru vineyards in Alsace, France, MacNeil examines "what, if any, limits should be set on a Grand cru's yield. Clearly, the stricter the requirements, the more impact and validity the designation Alsace Grand cru would have."[64] She continues: "A good property (producing Merlot in St. Emillion/Pomerol) may well do an additional summer prune to further restrict yield to make only, say, 3,000 quarts of wine for every 2.5 acres. A basic Merlot from California's Central Valley, France's Languedoc, or Italy's Veneto may have been made to about 3 or 4 times this yield, and the taste is obviously stretched accordingly."

Halliday and Johnson also note caveats with respect to the HYLQ relationship, including hail and rot. Yet, in general, these phrases, "the stricter the requirements, the more impact and validity," and "stretched accordingly," imply a direct dependence of quality that arises from increased concentrations of good flavors in fruit from lower-yielding grapevines—like that of curve 1 in figure 7. Claims that low yield is key to wine quality imply a relationship unlike curve 2. Some high-end wines in California carry back labels touting their extremely low yield, which implies a relationship consistent with curve 3 or possibly curve 1.

THE COMPLEXITY OF YIELD

Although the nature of the yield-quality relationship is fundamentally important to those in the business, there has been surprisingly little direct effort to resolve it. Crop yield in grapevine production depends on many factors over two seasons: varieties, weather, and cultural practices (as discussed in figs. 1 and 3). The propensity to initiate flower clusters and the number of flowers in a cluster varies among varieties, but also depends on environmental conditions, including but not limited to temperature, light on the developing bud, and vine water status. Fruit set (the fraction of flowers retained) of any variety is sensitive to environmental factors such as untimely rain and low temperatures. Vine row spacing, established at planting, affects yield rather directly. Once a vineyard is established, pruning takes place before the season begins, and shoot and cluster thinning during the season is the main means of regulating yield. Water and nutrient supply are usually regulated with the objectives of desirable vine growth and fruit quality, and therefore impact yield as well. Because the sources of yield are so varied, the task of determining the facts about HYLQ is enormous.

It is important to note that the HYLQ myth tells us that whether yield is reduced by winter pruning to lower shoots per vine, summer pruning (or cluster thinning), or poor fruit set, each action is said to have the same positive impact on flavor intensity. Thus, in the popular press, it is the yield per se that determines the fruit quality. If great winegrowing is as simple as getting as far to the left along curve 1 or 3 as possible, then it is hardly a challenge worth appreciating. In the following sections, we will turn to the empirical evidence related to various means of altering yield, keeping in mind that with so many paths to yield, it could turn out that none of the hypothetical scenarios for quality response to yield are a reliable generalization.

Variety

Wine drinkers know that Riesling wine tastes different than Chardonnay, that Cabernet Sauvignon tastes different than Merlot (but not as different as Riesling from Chardonnay), and that Gamay and Pinot noir wines are readily distinguishable from one another. A common viewpoint within the popular press on the role of varieties in winegrowing goes like this: "At the top level, estates are moving away from the varieties that yield oceans of cheap wine and concentrating instead on more flavor-packed varieties like . . ." [65] In other words, it is thought that varieties that are more fruitful make poor wine. But are these differences in varietal wine attributes due simply to differences in yield among those varieties, or is there something more at work? Klebs' Concept, described in the introduction, tells us that the traits of fruit arise from the variety and its interactions with the environment. At the very least, we cannot get a red wine by restricting yield of a white variety, nor realize the monoterpenes characteristic of Rieslings by restricting yield in Chardonnay. A grapevine variety's genotype is responsible for establishing flavor potentials, as well as yield potentials. The final fruit composition necessarily reflects the grapevine variety, as well as any impacts of the growing environment, including crop load.

Most popular winegrape varieties that exist today are here and sustained (in some cases for hundreds of years) because the resulting wines taste(d) good *and* because they are high yielding, when compared to the background vines from which they were selected. For all of these attributes to occur together, the varieties must also be adapted to their environments, so that they are sufficiently fruitful and ripen by the end of the season. In some cool climates with short growing seasons, winegrape varieties were probably selected for their early ripening. Pinot noir, for example, is early maturing, relatively low yielding, and seems to have a low temperature optimum for making color—a good combination in the relatively cool, short season of Burgundy.

Beyond the inherent yield and flavor potentials, differences in how varieties interact with the environment should be expected. There are many anecdotal claims about how varieties differ in their response to weather conditions. Pinot noir is unceasingly referred to as sensitive, difficult to grow, or with any number of synonyms. Karen MacNeil compares the putative yield responses of three popular winegrape varieties in the following way: "Cabernet Sauvignon and Chardonnay are relatively insensitive to yield in contrast to Pinot noir, which quickly loses its 'stuffing' at even moderate yields."[66] Thus, MacNeil says that Pinot noir has a lower yield threshold (or breaking point, as represented by curve 2 in fig. 7) for loss of quality than Cabernet Sauvignon or Chardonnay. This is a common opinion, but it would be interesting to learn whether there is evidence to support it.

Straightforward data on fruit and wines from vines that produced at various yields should not be hard to come by; however, the variety trials (where the plants are grown together in the same environment) that are so common in annual crops research and breeding programs are actually rare in winegrape research. Where variety trials have been planted, the resources to follow through with extensive measurements of fruit and winemaking have often been lacking. In fact, there are more and better comparative studies of leaf functions among grape varieties by research groups like Manuela Chaves's Instituto de Tecnologia Química e Biológica at the New University of Lisbon than there are of berry functions and composition.

There are more experiments comparing clones within a variety than there are variety trials. This probably reflects the fact that all the fruit in a clonal trial can be harvested and crushed together more readily than combining varieties, whereas variety comparisons generate additional expense and difficulties in establishing and maintaining vineyards with a mixed planting of varieties. The rarity of variety trials may also indicate disinterest in testing what is considered by some to be already known. Similarly, there is more investment in clonal selection within popular varieties than in breeding new varieties. According to one heavily cited review, clonal selection "has been used successfully in Germany and Burgundy to improve yields and maintain quality of Riesling and Pinot noir, respectively."[67] Thus, changing the genotype to increase yield in winegrapes is not inevitably tied to a loss of winegrape quality.

Pinot noir has been compared to other varieties in a few cases. In one study in northern Italy, Cabernet Sauvignon and Pinot noir vines were directly compared. When the vines were pruned to 30 or 50 buds/vine, "organoleptic tests showed that only in Sauvignon could the different treatments be detected in the taste of the wine. The panel tasters preferred the [Cabernet Sauvignon] wine from the 50 nodes/vine treatment."[68] Thus, when both Cabernet and Pinot were grown to very different amounts of crop in the same environment, the resulting Pinot wines could not be distinguished, but the Cabernet Sauvignon wines could, and in fact, the higher-yielding Cabernet Sauvignon was preferred. In another vineyard trial in northern Italy, Enrico Peterlunger and colleagues at the University of Udine studied Pinot noir grown on various training systems that produced yields from 7.5 to 9.7 t/ha and from 0.9 to 2.6 kg/vine. Their conclusion: "The training systems affected yield but showed little or no impact on grape and wine composition (sugars, grape and wine phenolics). Sensory analysis could not show relevant differences among training systems."[69]

Even when crop on Pinot noir vines was halved by cluster thinning in another study, only one or two out of twelve sensory attributes were distinguished by a taste panel in any of the several years of the experiment. Over multiple years of study, six out of thirty-one sensory tests showed differences, and no single flavor or aroma attribute differed consistently (in more than one season).[70] When Pinot noir in Tasmania, Australia, was pruned to 10, 20, 30, and 40 buds, color was least in the fruit from vines with the lowest yield[71]—the opposite of what one would predict in accordance with HYLQ. In addition to being less sensitive than the Cabernet Sauvignon in one study, Pinot noir was found in several well-designed and well-conducted studies to be relatively insensitive to yield when the fruit could be matured to the target sugar (Brix) concentrations.

Halliday singled out Shiraz as a particularly sensitive variety: "If limiting the yield is important in maximizing the potential of Chardonnay, it is quite critical with Shiraz."[72] In an important long-term study of pruning and irrigation in Shiraz during the 1970s, vines were pruned to 20, 40, 80, and 160 buds per vine, which had correspondingly large effects on yield:

> The 1975 wines were tasted by the Research Station panel when the wines were three months old. Each treatment was paired with a control wine, and the taster was asked to determine if there was any difference and to score the degree of difference on a preference scale of five points. In 1975, the wines from the NI (not irrigated) vines scored higher than the wines from the vines in the directional paired tasting, but pruning had no effect on the scores.[73]

> Contrary to popular belief, increasing the buds per vine, hence yield, generally increased the quality of these (wine color) parameters.[74]

That study is still a wealth of information today. Similar experiments conducted twenty-five years later produced similar results: 30, 60, and 120 buds per vine in Shiraz produced no significant difference in fruit color in the two seasons of the trial. The authors concluded:

> The current study has shown that higher crop loads can be carried by Shiraz grapevines grown in the Barossa region, causing only a delay in ripening and reduced sugar per berry, but with no deleterious effect on fruit composition as measured by pH, TA, or the concentration of anthocyanins and phenolics.[75]

Applying the HYLQ concept interchangeably among varieties ignores the biology of what makes one variety distinct from another. There is no biological reason or precedent in other crops that should lead us to assume that more yield in a variety is linked inversely and inescapably to lower concentrations of flavors in the fruit, when compared to a lower-yielding variety. For example, Petit Verdot is more fruitful (produces more clusters) than Malbec and many other red varieties, yet has much more red color in the fruit and wines. It is likely that varieties differ in their responses to crop load, but the limited empirical evidence so far does not support a high sensitivity in any particular variety. We like wines so much that we try to grow winegrapes everywhere, including locations to which some popular varieties may not be best suited. It is possible that part of what has been attributed to variety sensitivity to yield has as much or more to do with the variety interaction with the soil and weather aspects of the environment, about which there is still much to be discovered.

Vine Spacing

Voices of authority in the popular press, as well as a few viticulture consultants, hold that vine spacing should be dense for high-quality winegrowing. Dense spacing supposedly leads to small vines and low yields, similar to what was first common centuries ago in Champagne and Burgundy, and then in Bordeaux, and other parts of northern Europe today. Bordeaux wine expert James Lawther claims, "A large body of research has proved that an increase in the density of planting helps improve grape quality and, within limits, reduce yield."[76] Yet this statement, like many others in the popular press, stands alone with no supporting research cited that can be evaluated or reviewed. Some wine writers even go so far as to criticize those who do not acknowledge their wisdom.[77]

Vine spacing has varied over the history of winegrowing, from extremes of more than 15,000 vines/acre or more in Champagne, at a time when all the work was done by humans, to only a few hundred vines/acre in Almeria, Spain.[78] Even in antiquity, spacing varied greatly from region to region.[79] Row spacing was not a concern when vine growers layered shoots into the soil in a willy-nilly manner in order to replace diseased or dying vines. Later, vine rows shifted for practical reasons—to accommodate animals and then machines. Growers in various environments have no doubt arrived at spacing practices that work for their particular conditions.

If one begins with a single row on an acre and adds another row, the yield will approximately double, and the vine rows could be spaced far enough apart that there would be no interaction between them. As row spacing gets tighter, the grapevines interact more and grow less, and although the yield per plant is less, the total yield per acre is increased. Thus, in vineyards with higher plant densities, yield per acre is generally higher, not lower,[80] as eminent French viticulturist Louis Ravaz demonstrated in vineyards a century ago.[81] Indeed, closer plant spacing has been a significant contributor to yield increases in crop production throughout the United States in the twentieth century and elsewhere during the Green Revolution. This is the commonsense context of considering vine spacing—yield increases as rows are brought closer together from wider spacings.

There are caveats with respect to that relationship, however, including that there is a finite amount of both sunlight to be intercepted and soil to be mined. Adding rows captures more sunlight, turning it into grapes instead of letting it hit the ground—up to a point. As more rows are added, mutual (one row to the next) shading above ground increases, as does root interaction below ground. As the plant density increases and plants interact more (both above and below ground), the growth of each plant diminishes. At some point in increasing plant density, reproductive development (yield) stops increasing. This is well established in many crops, such as cotton, corn, soybeans, and olives,[82] but is usually only observed at extremely high plant densities. The same is true of grape-vines where, at higher vine densities, yields are unchanged as spacing between vines is reduced, and can even decrease with tighter spacing.[83]

This is the rationale that closer vines will be smaller and lower yield-ing, because of competition among vines for water and nutrients. "Increased vine density creates root competition and higher quality ultra-premium fruit,"[84] and "the greater the number of plants, the more the vines fight each other for nourishment, reducing the number of grape clusters per vine. All the vine's energy is channeled toward the clusters that remain, so boosting the concentration and quality of the juice."[85] Although plant biologists do not usually speak of "fights" or "channeling energy" in plants, there is good reason to expect that inter-vine competition for limited soil resources increases with closer spacing, similar to the above-ground competition for sunlight.

The point of diminishing yield will occur at closer spacings when resources are abundant, and at wider spacings if water or mineral nutri-ents are limiting. When supplemental water is unavailable, vine spacing is dictated by water availability compared to evaporative demand, in some instances becoming an important constraint, as is the case in the Canary Islands (fig. 8), where each vine is allowed a large volume of soil to mine for water, and foxholes are introduced in order to gather water and protect against the wind. There is some evidence that plant crowd-ing increases root depth and density, which would ameliorate the increased competition for water and nutrients to some extent.

One study on Pinot noir in South Africa found that closer spacing did result in drier vines when dry-farmed (that is, not irrigated).[86] That study was part of a series of spacing experiments that consistently showed the positive effects of earlier ripening and more color (often observed with moderate water deficits) under closer spacing. Another multiyear study combined row and in-row spacing treatments of Caber-net Sauvignon in Napa Valley, California. Yields were more than 20 percent greater in the narrow spacing treatment; however, no sen-

FIGURE 8. CANARY ISLAND GRAPEVINES IN THEIR FOXHOLES. Low soil water availability (and high, water-sucking winds) dictate the widely spaced and protected vineyard design. (Photo by Simone Diego Castellarin.)

sory differences could be detected in blind tastings of the wines made from vines at spacings of 3.1 × 3.1 feet, 3.1 × 3.6 feet, and 5 × 9 feet.[87] Why fruit color increased about 100 percent in the South African Pinot noir trials and did not change (or increased very little) in the several other spacing trials in Napa Valley is not known, but it may be that the fundamental cause of the enhanced color in the closer-spaced Pinot was the vine water deficit, rather than the vine spacing per se. The related "stressed vine makes the best wine" myth will get its time in the spotlight in chapter 3.

The consequences of in-row (vine to vine) and row (row to row) spacing of vines are the subject of much speculation and some research. Vines are usually trained and pruned to distribute shoots and fruit along the axis of the row. Because the vines form a continuous canopy of leaves along each row, regardless of the vine spacing, the effects of changing row spacing are greater than those of changing vine spacing within a row. This is especially true if the spacing of the shoots (buds left at pruning) along the row is unchanged, which is usually the case. The aboveground vine can be managed to develop a similar number of shoots and canopy size and shape, except for the number of vine trunks along the row. Hence, the vine (within-row) spacing need not affect much physiologically, and experiments in which in-row vine spacing was varied usually reveal limited effects on fruit and resultant wines. (When not trained along the row, like the vines in figure 8, in-row spacing follows the same principles as row spacing.) When planted far apart within the row, vines may not fill the row with canopy, leaving open and unproductive spaces. This again effectively leaves part of the land unfarmed. As this margin is surpassed, wider vine spacing will result in decreased yield, similar to the results of wider row spacings.[88] When yield is reduced simply by not farming part of the land, few effects are expected in the fruit.

Strangely, there are many more studies of in-row spacing than of row spacing. As in-row vine spacing decreased (vines were planted closer to one another), several studies found that there was no effect or that yield actually increased, rather than decreased as predicted by the popular wisdom. In an experiment on Riesling, the yield of the closest in-row spacing was significantly higher in one of five years, significantly lower in one of five years, and showed no difference in two of five years.[89]

Closer in-row spacing had no significant effect on anthocyanins or total phenolics in a study of Barbara over five years; increased color about 3 percent in Chancellor over a span of five years; had very little effect on any aspect of fruit composition and no effect on berry size or anthocyanins in Cabernet Sauvignon over a period of three years; and had no effect on terpenes (which produce flowery aromas) in Riesling fruit in three of four years.[90]

There is an abundance of vine trellis and training designs in viticulture. The most important characteristic of some trellis designs is the "divided canopy," which is somewhat similar to halving the row spacing. The number of buds/acre is increased up to 100 percent, and yields typically increase by almost 50 percent. In a study in which yields were increased by dividing canopies in the varieties Seyval and Chancellor in British Columbia, Andy Reynolds and colleagues made wines and conducted sensory analyses, concluding, "These data strongly suggest that high wine quality may be obtained from divided canopies despite large crop size and high crop loads." In a later review of the entire scientific literature on various winegrape training and trellising systems, Reynolds concluded the same is true in general—with proper management, high yields need not have detrimental effects on wine.[91]

Vine density, like yield, is regulated by law in many Old World vineyards. Some observers may associate regulations with knowledge of spacing optima for winegrape quality. That knowledge has not been established, however, and it may not be possible to do so, although some regions may have found spacing (and other cultural practices) that works for their environment and economic situation. Writer John Haeger investigated vine spacing in Pinot noir by visiting vineyards and in part by gaining access to informal in-house studies. The limited effects of spacing observed in those studies, combined with his own observations that tight vine spacing was present in vineyards that were the sources of both the best and significantly not-the-best Pinot wines in Burgundy, led Haeger to conclude that spacing was not a key factor in Pinot quality. A symposium at the 1999 annual meeting of the American Society for Enology and Viticulture, in which several spacing trials were discussed, found little evidence in the fruit and wines produced to compel a movement toward closer spacing.[92] Despite all this, American viticulturists' opinions on how to use vine spacing are decidedly mixed.[93]

There are important practical considerations for investment in the up-front costs of denser spacing: weighing the implications for efficiency in production; mechanization of vineyard operations; increased costs and earlier time to economic returns of planting more vines. There have also been some bad experiences with vines planted too densely for the climate and soil. In more verdant environments, closer row spacing will diminish wind and increase humidity. According to one grower, "If you wanted to invent a system designed to promote rot and mildew in California, it's high-density planting."[94]

In general, studies of in-row spacing have yielded results similar to those found in Reynolds's work with Riesling, from which he concluded, "These studies showed minimal impact of vine (in-row) spacing upon yields and fruit composition under situations where soil conditions are not limiting."[95] Thus, the appropriate generalization about in-row spacing based on published research is that it is unlikely to play a large role in determining yield or fruit composition. This can be expected as long as the vines are spaced close enough in the row to produce an adequate and continuous canopy along the row, and not so densely that barrenness, disease, or other problems develop. Vine row spacing must allow equipment to move between rows, and narrow spacing increases all vineyard costs while generally increasing yield.

Dense vine spacing is promoted by some wine journalists and in conjunction with low yields, implying that high-density planting is a means to achieving low yield and high quality. Although planting more densely to obtain lower yield and higher quality would be consistent with the HYLQ rationale, it most often leads to higher yield. It is possible that this received wisdom derives from no more than observing what is done currently in places where fine wine is produced. Where vine density is traditionally high, the environmental conditions typically make for slow growth and low yields; therefore growers plant vines closer together in order to farm all of their available land. Thus, rather than being planted close in order to obtain low yields, vines were planted close *because* of low yield. That was the reason for very close spacing in medieval Europe in the first place—to get as much yield out of the land as possible.

Vine spacing can affect the microclimate (environmental conditions around the clusters) that the buds and fruit respond to. However, the point of canopy management practices (including trellising, shoot positioning, shoot thinning, etc.) is to make the canopy adhere to microclimate objectives, which, when done properly, would ameliorate vine spacing effects.[96] The role of spacing, particularly in-row spacing, also remains unclear because yield is regulated by pruning and thinning. Perhaps accordingly, vine spacing is (and has been) largely a practical and economic matter.[97]

Pruning and Cluster Thinning

Pruning and cluster thinning are the most direct and common means for manipulating yield in winegrowing, and they can be adjusted each season. A third means is shoot thinning, although this technique is also used to manage the canopy microclimate. When HYLQ is raised in discussions of vineyard management, pruning and cluster thinning are usually the implied subjects. The HYLQ myth did not come into existence by mistake; there are studies dating back as far as 1904 containing results and interpretations consistent with HYLQ predictions, but the number and history of studies with contrary results are quite surprising.[98]

Investigations of putative yield effects on wine sensory attributes evidently began in the 1950s at UC Davis, long after the HYLQ myth had assumed dogma status. Professors Maynard Amerine, Robert Weaver, and others conducted field trials with varied pruning, cluster thinning, or both. They then measured vine yields, followed a consistent protocol to make the fruit into wine, and used the recently developed sensory methods to evaluate wine quality. Combining several vintages, their data show little or no response of wine scores to wide changes in yield (up to a fivefold range of yields) of Alicante, Carignane, Grenache, and Zinfandel (fig. 9A);[99] however, wine scores for Zinfandel may have decreased at high yields per vine (fig. 9A). In 1961, Amerine and Weaver reported on another set of experiments with Grenache and Carignane that appear to show more sensitivity of wine scores to yield (fig. 9B) but at much higher yields; the authors concluded, "The final wine rating was just as high in over-cropped Carignane vines as in normal-crop fruit, and the difference was only slightly in favor of the normal crop in Grenache."

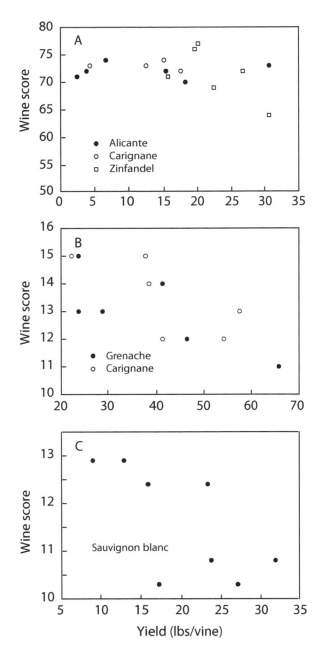

FIGURE 9. THE RESPONSE OF WINE QUALITY SCORES FROM EXPERT WINE PANELS TO GRAPEVINE YIELD. For (A) Alicante, Carignane, and Zinfandel grown in Davis, California, and harvested in 1949, 1950, and 1951 (Weaver et al. 1957), (B) Grenache and Carignane (unirrigated) grown in Davis, California, and harvested in 1957, 1958, and 1959 (Weaver et al. 1961), and (C) Sauvignon blanc grown in the Golan Heights, Israel, and harvested in 1992, 1993, and 1994 (Naor et al. 2002).

Brian Freeman and Mark Kliewer conducted studies in the early 1980s that manipulated yield via irrigation and pruning in both Shiraz and Carignane. The large study on Shiraz in Australia, in which severe pruning (i.e., dramatic reduction in yield) in fact *decreased* or had no effect on wine color in three of four years, has already been mentioned.[100] From their work together in California, Freeman and Kliewer concluded, "Contrary to popular belief, increasing the pruning level, hence yield, generally increased these quality parameters."[101] In the late 1980s, a study of Cabernet Sauvignon in Napa Valley by Kliewer at UC Davis found that taste panels could successfully distinguish between 6.6 and 11.0 tons/acre wines in only one of two years, and could not distinguish between wines made from vines with smaller differences in yield.[102]

In addition to several studies in the 1980s already mentioned, many studies over the next two decades reported similar observations. In Pinot noir, large differences in yield (twofold) caused no differences in fruit color or pH,[103] and cluster thinning to reduce yield about 35 percent had no effect on skin tannin or color in a study of Merlot in three consecutive years. Other Australian studies in hot regions reported a nil to weak relationship between yield and berry color in Shiraz and Cabernet Sauvignon vineyards. In yet another comprehensive study of Shiraz wines produced in South Australia, John Gray and coauthors were unable to find a relationship between yield and their "wine value index" across a large sample of growers in several regions.[104]

Leading grapevine scholar Markus Keller and colleagues conducted a series of studies in Washington State that involved changing yield in various ways. In one five-year study, after reducing clusters by about 25 percent in Cabernet Sauvignon, Riesling, and Chenin blanc, Keller and colleagues found that "cluster thinning and its timing had little or no influence on shoot growth, leaf area, pruning weight, berry number, berry weight, and fruit composition (soluble solids, titratable acidity, pH, color) in both the current and subsequent seasons." In another five-year study, reducing clusters by about 40 percent had no effect on fruit composition (including color) in Cabernet Sauvignon. In 2010, Keller and colleagues conducted novel experiments in which the temperature of buds was increased in order to enhance their fruitfulness, and found that "although yield per shoot varied threefold among treatments, differences in fruit composition were minor."[105] Yet the world of winegrowing has not assimilated these many experimental contradictions of HYLQ.

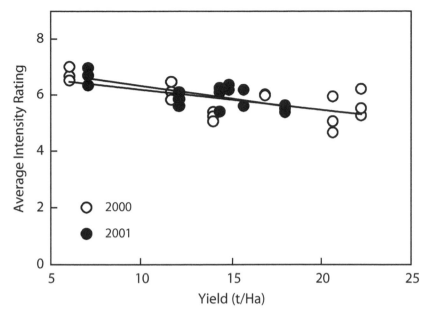

FIGURE 10. ASTRINGENCY OF CABERNET SAUVIGNON WINES MADE FROM VINES WITH VARIOUS YIELDS IMPOSED BY PRUNING (Chapman et al. 2004a). The coefficient of determination (r^2) indicates how well the data fit a direct, linear relationship, and can be interpreted as indicating how much of the variation in astringency is associated with changes in yield.

If the data fit a straight line perfectly, r^2 would be 1.0. For these data, r^2 is 0.41, which is not a strong correlation; however, the relationship is significant at the level of 95 percent (in other words, there's only a 5 percent chance that the relationship reflected here is due to chance alone). These values indicate that although less than half of the variation in astringency was attributable to a linear relation with yield, there is a high probability that that astringency is related to yield. The r^2 is dependent on several contributing factors—including the sample number; it is possible that the relationship would have been stronger with more data.

There are a few authors, however, who have reported positive changes in fruit or wines (such as increased fruity aroma) with increasing yield, regardless of harvest criteria.[106] For fruit harvested at similar Brix, the consequences of yield reduction by pruning or cluster thinning for grapes and wine, when present, are often like the increased astringency in Cabernet Sauvignon wines shown in figure 10, changes that are sometimes statistically significant but also small.[107] Note that the wine astringency ratings, which are closely correlated with the tannin concentrations, varied linearly (as in curve 1) but only from about 5 to not quite 7 on a scale of 1 to 10, while the corresponding differences in yield varied more than threefold.

YIELD AND RIPENING

To be sure, there are studies showing negative relationships of color or other quality attributes to yield, as predicted by HYLQ. For example, Sauvignon blanc wine quality scores were highest at the lowest yield and lowest at the highest yield in a study that included some (but not all) fruit harvested at the same Brix (fig. 9C).[108] The maturity (Brix) of the fruit at harvest is important for understanding the role of yield in winegrape quality, because most viticulture experiments harvest all fruit on the same day, but the most common fruit response to yield reduction by pruning and cluster thinning is advanced fruit ripening.[109] Thus, fruit reach ripeness sufficient for winemaking earlier when lower crop loads are carried, and they are riper at a given harvest date. This creates a problem for interpreting the results of those studies, unless the only quality issue is advancement of ripening.

In experiments with Shiraz in Australia's Barossa Valley, researchers Tony Wolfe, Peter Dry, and colleagues found a consistently negative relationship between fruit anthocyanins and crop yield in three different seasons (color decreased as yield increased).[110] In this study, however, sugar accumulation lagged in the higher-yielding treatments, making it difficult to know how much of the differences in color were due to differences in fruit maturity. When the fruit of vines with significantly different yields are harvested on the same date, wines made from unripe (higher-yielding) grapes are often found lacking.[111] For example, in one early study when crop was thinned, the intensity of "good wine aroma" was dramatically increased, but the higher cropped vines had less mature fruit at harvest.[112] This is consistent with HYLQ, but the aromas probably reflected riper fruit at lower crop loads rather than an effect of yield per se.

In most cases when yields varied but fruit were harvested at the same Brix (rather than on the same date), studies have found little or no significant differences or loss of quality attributes in fruit or wine (table 1).[113] However, my analysis is based on what I gleaned from studies conducted for purposes other than testing HYLQ, and it is notoriously difficult to harvest fruit at precisely the desired Brix. This issue should be addressed experimentally, and that work should include harvesting fruit based on an experienced winemaker's sense of ripeness, in addition to harvesting on the basis of Brix.

Furthermore, for growers who do not have the luxury of waiting until later in the season, getting to a higher sugar concentration is tantamount to quality. For these growers, a "vintage year" was a year

TABLE I   SAMPLE OF RESEARCH STUDIES THAT REPORT LITTLE OR NO EFFECT OF
CHANGES IN YIELD ON FRUIT OR WINE ATTRIBUTES

	Pruning	Cluster thinning	Spacing and trellising
Cabernet Sauvignon	✓	✓	
Chardonnay Musque		✓	
Merlot		✓	
Pinot gris	✓		
Pinot noir	✓		✓
Riesling	✓	✓	✓
Shiraz	✓		
Seyval blanc			✓
Zinfandel		✓	

NOTE: Most of the studies harvested fruit at the same Brix. The studies are sorted by the means
employed to vary yield.

when the sugar concentration reached an adequate value for winemaking at all, and Chaptalization was not required.[114] In the milder climates of southern Europe, California, and other warm regions, every year is a vintage year in this regard, and growers in these accommodating climates can simply wait a little longer and harvest a higher crop load at the same ripeness. Indeed, longer "hang times" have become associated with desirable wine flavors for some (but not all) wine experts.[115]

To many both inside and outside the realm of viticulture research, it comes as a surprise that for more than half a century, pruning and thinning studies have reported small or undetectable effects on fruit and wine. While it is widely recognized that increased yield has one major effect in delaying ripening, the possibility that higher yield could be a means of achieving better flavors derived from longer hang times (a longer ripening period) seems to go unrecognized. Yet many continue to consider higher yield, however it is attained, a problem that dilutes the fruit.

THE DILUTION EFFECT

Great wines are usually described as having concentrated flavors (although there are experts who feel that this has been taken too far). Almost without exception, the dilution concept is invoked by wine critics to explain how both high yields and large berries cause low-quality fruit and wines. According to Robert Parker, "High yields from perennial overcroppers result in diluted wine."[116] Although that may sound feasible to many, it sounds suspicious to this plant physiologist.

In a strict sense, to dilute is to add solvent (water) to a solution, such as a wine. There is no change in the number of solute (e.g., flavor) molecules, there is just more water.[117] Adding water to a wine dilutes the wine and decreases the concentration of all dissolved solutes by increasing the volume of solvent. In a more casual sense, we can think of a berry or a wine with a low concentration of a flavor as being dilute compared to another with a high concentration. Again in a strict sense, if high yields dilute fruit and resultant wine, then higher yields mean that the "higher" part of high yield is more water. This thinking may be intuitive for some, but it is a phenomenon that does not occur in plants.

Consider one example from a study of strawberry varieties that compared a number of traits. In the findings, yield did not correlate with the two quality traits that were extensively measured (Brix and color) among about 100 varieties that ranged in yield from 400 to 1,500 grams of fruit per plant.[118] Another example can be found in average tomato yields in the United States, which have increased eightfold when compared to the 1940s. The increase has been pretty steady, and there are a number of factors involved—genetics is a big one, but there have also been a variety of horticultural improvements. The increase in tomato yields was not accompanied by a dilution in tomato fruit sugars—in fact, the content of soluble solids has stayed constant or gone up a bit.[119] Similarly, winegrapes are (or in many cases can be) harvested at the same sugar concentration independent of their yield. Indeed, the historical objective of reducing yield was to reach a sugar concentration by the end of the season. The dilution explanation is not a part of agronomic or horticulture crop understanding, probably because it misses important physiological facts of the reproductive development that leads to fruit size and yield, and implies one or more false premises.

Wine writer Rod Phillips provides the intuitive sense of how lowering yield would lead to higher concentrations in winegrapes: "In general terms, the lower the yield, the more flavorful and complex the wines. Yields are often reduced by 'green harvesting,' which involves picking (and throwing away) a proportion of the bunches of grapes on each vine before they begin to ripen. This allows the smaller number of bunches remaining to benefit from all the nutrients the vine absorbs."[120] This implies that there is a fixed amount of flavor to be made by grapevines in a vineyard, and the variable is the number of bunches into which that flavor is distributed, but in general, as we noted with strawberries and tomatoes, the concentrations of solutes in mature fruit are relatively stable among yields. Higher yields are accompanied by higher amounts of sugar, tannin, anthocyanins, and so on to be removed with the crop. One version of "the dilution effect" theory suggests that the real issue is yield per plant, rather than yield per unit land area. If the yield per plant is less, the quality of fruit and resultant wines is better, as spelled out by wine writer Oz Clarke: "If you produce 50 hectoliters/hectare of wine from 3,000 vines, each vine will produce four times as much fruit as if you've planted 12,000 vines. So the juice will be much more diluted."[121] However, this is confused thinking that assumes that each plant has the same amount of flavor to deliver to its fruit, regardless of its size.

It is not clear how high-yield dilution would take place in order to make thin wines, as higher yield is not a matter of simply adding more water to the fruit. Berries and other fruit grow and ripen under a genetically controlled program. This growth program takes the fruit (and seed) from bloom to maturity, and has characteristic traits with regard to the concentrations of various solutes at maturity. Growth involves water import into cells, but that import is tied to solute concentrations inside the cells. The synthesis of some flavors keeps pace with water influx associated with growth, while the synthesis of other flavor compounds does not. If dilution is present in the vineyard, then we would expect the concentration of all solutes to decline similarly as the water moved in, but this is not what happens. Indeed, the accumulation of color, fruity aromas, and other ripening-related compounds is necessarily the opposite of dilution—more of these compounds are made than the amount of solvent water that is imported, causing their concentrations to increase—and it is these increases that define ripening. Some solutes, such as malate and the strong veggy aroma compound MIBP (a methoxypryazine), are lost by various metabolic means—their concentration decreases in ways unrelated to water content. Thus, the amount of a flavor solute in the berry increases when solute synthesis in the berry exceeds its breakdown.[122]

To put the dilution idea in the HYLQ and BBB myths into a wider context, consider the increase in size of other crops, fruits, and plant parts. For example, fruit are modified leaves,[123] and leaves do not become less concentrated when there are more leaves.[124] And leaves are not less concentrated when the leaf is bigger, even though the concentration of some solutes can vary considerably. For example, when basil was grown in different light conditions that affected leaf growth, the concentration of eugenol was higher in leaves that actually *increased* in size (fig. 11).[125] Eugenol and linalool contribute to the aromas of both basil and wines. In that same basil study, the concentration of linalool

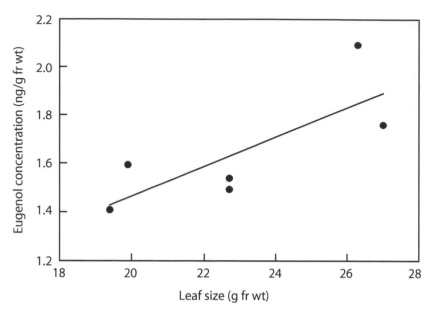

FIGURE 11. CONCENTRATION OF THE AROMATIC COMPOUND EUGENOL IN BASIL LEAVES OF VARIOUS SIZES. (Data from Loughrin and Kasperbauer 2001.)

responded to light environments, but its concentration was not related to leaf size. In basil leaves, light affected growth *and* the synthesis of several aromatic compounds, but in different ways. This is what should be expected, and therefore, should be the context for considering the impact of various environments on grapes and wines as well. Leaves are not less concentrated when larger; the concentrations of most solutes in mature leaves are relatively stable, regardless of leaf size. Environmental conditions, however, can and do alter development in ways that affect the concentrations of solutes in specific, environmentally sensitive pathways.

There *is* a "growth dilution" phenomenon in agronomy and horticulture that is associated with the correction of extreme nutrient deficiencies and their corresponding very low yields. When a severely deficient nutrient is supplied to the plant, the new growth response can be great enough that the concentration of some nutrients decreases as the plant grows faster than the nutrients can be taken up from the soil.[126] This growth dilution effect has most often been reported in measurements of leaf composition (leaves are commonly used to assay crop plants for their nutrient status and needs). Yet even there, the concentrations do not decrease uniformly across all solutes, which indicates that there is more going on than the simple influx of water to cause leaves to grow more. When grapevines were grown for several years at low nutrient supply and then supplied with nitrogen, both crop yield *and* the nitrogen concentration in the fruit increased, rather than a yield increase and nitrogen concentration decrease, as predicted by the dilution-based HYLQ theory.[127] Thus, the concentrations of mineral nutrients are dependent on physiology in complex ways, and are not the simple consequence of water uptake. It is important to point out that the growth dilution in mineral nutrition is not what the popular press and winemakers are referring to with respect to yield, dilution, and berry size.

There are a couple of other ways that berries might experience dilution/concentration. Plant tissues act as capacitors for water, charging up with water at night and discharging water during the day. Thus, there is a daily dilution and concentration that takes place in leaves, stems, and berries. Fruit harvested in the afternoon will contain less water than the same fruit harvested at dawn. In a bit of irony, night harvesting, popular in California for practical reasons (it's less dusty and cooler for the harvesters at night), takes fruit at their lowest concentration point on the daily cycle. The effect of night harvesting on flavor concentrations, however, should be very little, because daily shrinking and swelling, present before ripening begins, is almost eliminated in ripening berries.[128]

There are other scenarios in which the fruit can become concentrated by the loss of water. Fruit are harvested and partially dried in the making of Italian Amerone and other similar-style wines, and there have been some promising results in preliminary research that employs cutting the shoots to induce drying of the fruit on the vines in order to increase the flavor and color of red wines.[129] Another situation is the decision to simply wait to harvest until the berry's fresh weight diminishes. Although all solutes are concentrated as the berry size diminishes, these means of partial desiccation may do more than cause concentration due to water loss, because there are specific metabolic changes induced in the drying of grapes and peaches.[130]

A NOTE ON IRRIGATION

Some in the wine press promote the traditional myth that irrigation makes for bad wine, with the further assumption that the putative loss of quality is due to an increase in yield. See *The Wine Bible*, for example:

A larger yield, however, would result in thinner, lower-quality wine, which is why irrigation is prohibited [in Europe].

When the yield is high enough, the outcome is thin, bland wine.[131]

It is often assumed that where irrigation has been outlawed, it was done in order to control quality. It was during periods of overproduction in the 1930s and again in the 1950s, however, that bans on irrigation came into existence in France, as parts of regulatory packages designed to limit production—the original reason for restricting irrigation was to limit supply in the face of overproduction and low prices.[132] Although given the guise of quality control, a negative effect of irrigation on fruit or wine quality mediated by yield had not been clearly established or even investigated; what *was* clear was that the amount of wine being produced had to be reduced.

Claims that late rain or irrigation activities dilute the fruit are thus unfounded.[133] In 1979, Bryan Coombe, an Australian plant physiologist (who probably has contributed the most to today's understanding of grape berry growth and development), tried to dilute berries with late-season irrigations, and the results showed that the fruit didn't seem to notice his efforts.[134] We have some insight into how and why that could be. Once fruit ripening has begun, the transport of water, sugar, amino acids, plant hormones, and so on to the berry is almost exclusively via the phloem from the leaves, and not via the xylem conduits from the roots.[135] This seems to be the case to varying degrees in other fruit, including ripening tomato, mango, kiwi, and apple. In this way, the fruit appear to become hydraulically isolated from the soil water as ripening progresses.

In addition to wine writers and critics, some academics have also equated their observations on irrigation with the HYLQ concept, noting, for example, that "a grower aims for a balance between fruit quality and yield through conservative irrigation practices."[136] However, until relatively recently there has been surprisingly little direct investigation into the consequences of irrigation with regard to wine flavor, and this attitude is perhaps an example of the myth being passively received by academics. An analysis of all the studies on irrigation in grapes showed a relative dearth of wine sensory work related to irrigation (table 2.)[137] The results show that the number of studies involving irrigation is high, but that their prevalence decreases by orders of magnitude as the search is restricted to "irrigation and wine," and then to "irrigation and wine and sensory."

TABLE 2 FREQUENCY OF STUDIES ON THE EFFECT OF IRRIGATION OF GRAPES, THOSE CARRIED THROUGH TO WINES, AND THOSE THAT INCLUDE IRRIGATION OF GRAPES, CARRY THROUGH TO WINES, AND INVOLVE SENSORY ANALYSIS OF THE WINES

On the topic	Number of research papers	
	As of 2000	As of 2012
Irrigation and grapes	400	732
Irrigation, grapes, and wine	70	173
Irrigation, grapes, wine, and sensory	3	11

SOURCE: Data from the CAB Abstracts database.

Maynard Amerine (along with mathematician Ed Roessler) pioneered the science of wine tasting by developing the first rigorous sensory methodologies for tasting.[138] Some of the sensory tests were hedonistic (preference) tests, which can be challenging to interpret because the sense of quality is difficult to make objective or consistent among the judges. Difference tests are more straightforward and, with properly designed vineyard experiments, can be powerful in resolving how much of a vineyard effect is required in order to be detectable in the resultant wine. Considerably more sophisticated testing has been developed that is broadly called descriptive analysis.[139] In these experiments, the group of judges gets together and, based on blindly tasting the wines, decides which sensory attributes seem to be important or different. The judges then train with standards in order to align their sensibilities for those attributes, and then wines are rated in blind tastings for the characteristics that were previously defined. Descriptive analysis techniques inform about attributes, such as bitterness or "black cherry" intensity, and can be used to characterize sensory differences among wines across multiple attributes, with judges whose backgrounds are selected for uniformity and whose concepts of the sensory attributes are aligned in separate training sessions.

My studies with sensory scientists depended in part on wine scores obtained by difference tests, in which judges were asked to show that they could consistently distinguish between wines made from vines that received high or low irrigation; and later, my research utilized descriptive analysis. Other studies—for example, those by Ben-Ami Bravdo's group—used the so-called Davis 20-Point System, which has been criticized for its alleged emphasis on wine faults. Nevertheless, all kinds of sensory measurements can be helpful, including preference (quality scores, such as the Davis 20-Point System or the one used by your favorite wine critic), difference tests, and descriptive analysis methods. An advantage of difference testing and descriptive analysis methods is that the information generated does not rely on the judge's personal preferences, just on the taste and aroma intensities. Information from difference tests indicates whether various vineyard practices are effective, but the information about specific sensory attributes is of greater utility than difference tests for vineyard management and winemaking decisions on wine style.

Experiments are more quickly and easily completed when chemistry is used to analyze fruit composition; continuing on to winemaking and analysis of wine composition is fraught with difficulties, and then to persevere to evaluate the connections to the human taste experience has been uncommon. Although the wine taste experience is the ultimate concern, from the time of Amerine's earliest sensory work, reliable alternatives to the use of difficult human subjects have been sought. Only recently have methods of measuring large numbers of flavor and aroma compounds been developed, and it will require considerably more research before changes in those compounds, such as the floral-scented damascenone, can be interpreted with respect to the wine sensory experience. Sensory analysis remains essential in addressing the consequences of most vineyard parameters for wine flavor.

Although there were a few significant studies earlier, progress on irrigation, winegrape water relations, and how they impact wines truly commenced in the 1980s. In the previously mentioned studies by Freeman and Kliewer, changes in fruit and wines of Shiraz and Carignane were found to be due more to irrigation than to the large yield changes from pruning treatments. Ben-Ami Bravdo and colleagues applied deficit-irrigation treatments to Cabernet Sauvignon and Carignane grown in Israel, and interpreted their data as indicating a relationship between lower quality and higher yield—but the data show that decreased irrigation amount actually improved Cabernet Sauvignon wine quality, independent of yield.[140]

At about this same time (1983), I was hired at UC Davis to work on irrigation and mineral nutrition. Water use was a looming environmental issue, for which growers needed more information on how water applications or water deficits affected vines and wines, in order to develop prudent irrigation plans. At that time, only a few irrigation studies had been published, and none from work done in premium winegrowing regions of Europe or the New World, such as Napa

TABLE 3   WINE AND FRUIT RESPONSES TO EARLY AND LATE DEFICIT IRRIGATION
TREATMENTS

	Δ Color + flavor	Δ Yield	Δ Size	Δ Phenols	Δ Anthocyanins
Continual vs. early deficit	✓	↓	↓	↑	↑
Continual vs. late deficit	✓	↓	↓	↑	↑
Late deficit vs. early deficit	✓	↓	↓	↑	↑

✓ sensory evaluation distinguishable
↓ drop
↑ increase
NOTE: Irrigation was withheld before the fruit-ripening period in early deficit and during the fruit-ripening period in late deficit. All fruit for these wines were harvested at a similar Brix.

Valley. In our first study, we withheld water from drip-irrigated, hillside Cabernet franc vines on the eastern side of Napa Valley, either before or during ripening. Importantly, wines were distinguishable based on when in the season the vines experienced water deficits (as shown in table 3).[141] A common reaction when I presented these results was to attribute the changes in fruits and wines to the lower yields (and smaller berries)—the dilution effect.

I was never comfortable with the implication that the only aspect of plant metabolism that was sensitive to water availability was water uptake and movement into fruit. During my graduate student days, there was much excitement in the world of plant water relations about "osmotic adjustment," the beneficial and active accumulation (rather than dilution) of some "good" solutes in response to water deficits. At that time (in the early 1980s) specific metabolic pathways were being revealed as sensitive to water status. This notion from the world of wine, that in the face of water deficits everything stayed the same metabolically except there was less water sent to fruit, just didn't jibe with plant biology.

In addition to the difficulties of quantifying wine and fruit quality, ascertaining how yield affects those wines is further complicated because the paths to high and low yield are diverse, and irrigation is just one of the many potential variables. Although the generalization is true that crops produce more with more available water, the details and the exceptions to the generalization are extremely important. In one of the first reviews of grapevine water relations, Smart cited studies in which irrigation both increased and decreased grape yield.[142] We will see that in addition to the roles that water availability plays in yield, most other aspects of grapevine growth and development are affected by water, often in ways that impact fruit composition.

COMBINING YIELD CONTROL PRACTICES TO EVALUATE HYLQ

We have already noted how the popular wine press tends to assume all sources of yield control—spacing, pruning, cluster thinning, low fruit set, and so on—follow the HYLQ concept. This is only appropriate if it is yield that is the fundamental driver of fruit composition and subsequent wine sensory attributes. We see the correlation as predicted by HYLQ in comparisons of, for example, yield and wine quality for wine-growing regions moving south in both France and California. However, the cause of those correlations is not known, and it seems unlikely that yield is the primary culprit, given how easy it is to prune, cluster thin, and reduce inputs to lower yields. The power of the experiment in testing causation is shown when the predictions of HYLQ (and BBB) are assessed by managing grapevines in different ways to attain low yield or small berries.

In order to investigate this concept, Jean-Xavier Guinard and I, with our graduate student Dawn Chapman at UC Davis, designed a fairly direct test of the HYLQ hypothesis, in which we changed yield by employing the three most common cultural practices employed in wine-growing: pruning, cluster thinning, and irrigation management. With six pruning and six cluster-thinning treatments, yield of Cabernet Sauvignon in Oakville, California, was adjusted over a threefold range from about 6 to 20 t/ha. The fruit were harvested on different dates but at the same ripeness (as indicated by Brix), made into wine under the same conditions, and subjected to chemical and sensory analyses using the descriptive analysis methodology. A panel of experienced wine drinkers evaluated aromas, tastes, and mouth feel attributes of the wines, and chose terms such as bitterness, astringency, veggy aroma, red fruit flavor, and fruity aroma to describe the resultant wines.

There were significant differences in nine or ten of the fifteen sensory attributes when yield was changed by pruning. At high yield, tannin concentration was low (see fig. 10). As one might predict from HYLQ, the concentration of tannin was inversely related to yield; however, there were changes in other attributes that were not consistent with the HYLQ concept. Wines from vines pruned to higher yields were *less* veggy and *more* fruity, both considered to be positive responses.[143] When the same panel of judges assessed wines made from vines with a similar range of yields, but in this case imposed by cluster thinning, they found only three of fifteen sensory attributes were affected, similar to several other studies with cluster thinning.[144] Something was different about how the fruit developed when yield was changed by pruning rather than by cluster thinning at veraison.

In a companion study, Cabernet Sauvignon in Oakville, California, was subjected to three irrigation treatments: minimal irrigation (essentially no irrigation), standard irrigation (the rate of water application that growers typically use), and double irrigation (twice the rate that growers normally use). These irrigation regimes caused the vines to experience different degrees of water deficits during ripening and to produce yields that varied from 15.0 to 21.7 t/ha—a much smaller range of yields than imposed by pruning or cluster thinning.[145] Nevertheless, ten wine sensory attributes were affected by the irrigation treatments. The lowest-yielding minimal-irrigation-treatment wines were rated significantly *higher* in vegetal aroma, bell pepper aroma, and astringency. Water deficits (with lower yields) led to wines with *more* fruity and *less* vegetal aromas and flavors than wines made from vines with high water status. The responses of the veggy and fruit characteristics in the irrigation experiment were the inverse of the yield relationships for yield adjusted by pruning.

The decrease in veggy aroma and flavor at higher yields in the Oakville Cabernet Sauvignon pruning trial was surprising, but Andy Reynolds reported a decade earlier that cluster thinning had similarly increased some veggy aromas in Pinot noir. Prior to our work on yield at UC Davis that was released beginning in 2004, Reynolds's work was the only study that varied yield and evaluated the resultant wine-sensory attributes using descriptive analysis methods, in order to identify sensory attributes that responded to treatments. In that study, Pinot noir grown in Oregon and British Columbia was pruned to have high or low yield (100 percent and 50 percent crop load, respectively).[146] The wines from the high-yielding vines produced less "grassy" but more "bell pepper" aroma, and no differences in flavor. The conventional wisdom only deals with concentrating the good stuff, while less thought is given to the role of undesirable solutes in the HYLQ or dilution discussion. Some kind of a vegetal flavor/aroma used to be considered part of the varietal character of Cabernet Sauvignon, but in the past twenty years, many producers and critics have tried to get away from all vegetal characters, and away from MIBP in particular. The Oakville Cabernet Sauvignon pruning trial showed that higher yields reduced the amount of MIBP—and therefore reduced the veggy characteristics of the wines.[147] Note that this *is* consistent with the dilution idea; it just happens to occur with a compound that is considered undesirable by some producers and consumers.[148]

These varied wine-sensory profiles simply cannot be explained by a single mechanism in the vines that depends on crop load, and thus cannot be explained by the HYLQ paradigm. There is something about pruning, cluster thinning, and water deficits that is important to the grapevine, fruit, and wines made from those fruit. For example, water deficits are known to increase the stress hormone ABA, which directly activates the synthesis of the red-colored anthocyanins, but pruning or cluster thinning to lower yield has no known effect on ABA. Also, the effect of cluster thinning on the microclimate of the remaining clusters would be small, whereas the cluster microclimate must have been different when vines were pruned to have so many different shoots in the pruning treatments. If the high yield was the overriding issue in the grapevines, the flavors of the wines from all methods would have nevertheless converged around the causal changes in crop load.

BERRY SIZE

As with HYLQ, all paths to smaller berries are invoked in the wine press as means to better wines. For example, most of the possibilities in a vineyard setting that might affect fruit size—variety (or clone), soil environment, and aerial environment for both reds and whites—are summoned within wine descriptions from Napa-based Hess Collection: "The Zinfandel in this blend comes from an old clone grown in Dry Creek Valley. We chose this clone because of its small berries that deliver great color and intense fruit character." The sandy clay in the vineyard gives rise to small berries and intense fruit flavors for some Hess Collection Napa Chardonnay wines. They also produce small berries in their Argentine vineyards, where the climate conditions "produce smaller more concentrated fruit."[149]

Should we expect smaller berries to have the more concentrated solutes as a general rule? An important consideration here, similar to the traditional thinking about yield, is whether it is the berry size per se that is a key to fine winegrowing. We can take the same approach in evaluating the BBB myth as with HYLQ, by considering the empirical evidence for some of the ways that berry size is affected.

Embedded in the BBB myth is the dilution idea, discussed earlier in some detail. The "dilution effect" in viticulture refers to flavor compounds such as malate and eugenol. The BBB myth implies that there is a fixed amount of these flavors in each berry and that the variable in berry size is the water. If there is in fact a fixed amount of sugar and flavor for each berry regardless of the variety, and the variable is only the water involved in making the berry size, then yes, smaller berries will have higher concentrations of solutes. But we already know that this is not how the flavor content or the size of berries is established.

Perhaps the situation with the organic acid tartrate most closely approximates the simple dilution idea, in that tartrate is synthesized before the onset of ripening, and the additional berry growth during ripening effectively dilutes the tartrate. The delayed ripening associated with higher yields may also contribute to lower concentrations of some solutes on a specific date, but this is not dilution. On the other hand, an important corollary of the BBB myth takes into account the undisputed spherical geometry of the berry and the localization of some important solutes within the berry. Anyone who cuts open a red berry can see that the berry is round and that the color is in the skin. Color, as well particular tannins and other flavor compounds, accumulates in the thick-walled skin cells rather than in the flesh of the berry.[150] Upon crushing for fermentation, there develops an inescapable dilution of those solutes in the skin by nonskin sap of the flesh. These simple facts have led academics, producers, and wine writers to largely accept the BBB paradigm in most vineyard scenarios.

Variety

By 1826, an encyclopedia of agriculture stated: "Small berries and a harsh flavor are universally preferred for winemaking."[151] Several other horticultural books of that era make similar statements, referring to fruit that are indifferent or harsh for the table but are good for winemaking.[152] At the turn of the twentieth century, Thomas Munson, known as the Father of American Viticulture, reported: "It is found that small berry species generally possess properties for wine making far superior to the large berry species hence if one seeks to produce varieties for wine making he should not neglect those with small berries."[153] Although it was sound advice to not just go for the big-berried varieties, these statements give pause. First, just a little searching online reveals that exactly the same sentences or phrases appear sans attribution in multiple publications of the era. Thus, the frequency of the statement at the time may not reflect a considered opinion as much as a comment reiterated, much as Wikipedia content is reproduced on numerous other websites today. Second, the early references to desirability of small berries were provided in the setting of comparisons that included table grape varieties and alternative grapevine species; small and harsh were perhaps preferred for winegrowing in this context. Early European settlers in North America were impressed with the abundance of native grapevines, and therefore expected to have an easy time making good wine. Unfortunately for the settlers, the species present in North America were different from those back home.[154] Thus, for early American winegrowing, Munson may just as well have said, "Go for the small-berried species used in Europe, *Vitis vinifera,* which makes the best wine compared to those that are native to the United States."

Winegrape and table grape varieties are significantly different in their genetic makeup. The early classification of varieties into groups went like this: *occidentalis,* the small-berried winegrapes of Central and Western Europe; *orientalis,* the large-berried table grapes of Central Europe to West Asia; and *pontica,* the intermediate varieties from Eastern Europe and countries around the Black Sea.[155] The intermediate types of grapes are correspondingly intermediate in their genetic makeup.[156] These genetic differences reflect human selection that was different for table grapes and winegrapes, but it is likely that all cultivated grapes reflect an overall selection for bigger and sweeter berries than their native progenitors.

If winegrape quality was simply a matter of growing smaller-berried species, this feat would have been accomplished with native North American species long ago—relieving us of the need to graft *Vitis vinifera* scions onto rootstocks. Table grapes (like most fresh fruit) are prized for size. There has been more breeding conducted for table grapes than for winegrapes, much of it done directly for size, with breeders taking size manipulations about as far as they can. However, there are plenty of small-berried grapevines that make unpalatable wines.

Nevertheless, the academics have been largely in agreement with the BBB myth. At UC Davis, Albert Winkler and Maynard Amerine stated in 1943: "Although ordinary wines can be prepared from varieties with a large berry, the better flavored wines require a smaller berry size because of the more favorable surface-volume relation for extracting flavor and color from the skins."[157] This exact statement, which boldly says that the genetic basis of berry size alone limits the potential wine quality, appeared in wine books by William Cruess and Amerine that were published from 1941 until 1980, apparently without the benefit of any measurements or validating experiments. UC Davis enologist Vernon Singleton restricted the BBB paradigm to red varieties, but to him it was again straightforward: "Since smaller berries would have more skin surface per unit of berry weight or volume, it seems clear that red pigment concentration and perhaps therefore red wine quality would be higher with smaller berries that are otherwise similar."[158]

As noted, there are surprisingly few studies that compare varieties head-to-head within any given environment. In perhaps the earliest study to address berry size differences among varieties (conducted in 1966), Singleton reported the average berry size and total phenolics (which include tannins and anthocyanins that accumulate in the skin) in twelve varieties, including red and whites.[159] An analysis of the data shows a significant but rather weak negative relationship ($r^2 = 0.61$) between berry size and the concentration of total phenolics—that is, lower concentration in bigger berries. This is loosely consistent with the BBB concept, but nowhere close to the size effect predicted by changes in the surface-volume ratios (i.e., as shown by curve 3 in fig. 7).

Variety comparisons conducted since have seldom found a correlation as strong as Singleton's relationship between fruit size and composition for any solute. When berries have been harvested from different varieties at the same time, the concentration of sugars was sometimes higher in smaller-berried varieties, but often unrelated to berry size.[160] In large variety trials with thirteen and twenty-three varieties, berry size had no relation to the sugar concentration (Brix) of the harvested fruit[161] (data for the thirteen variety trials plotted in fig. 12B). Similarly, in each large study, berry size and the concentration of organic acids lacked correlation. No correlation was found between berry size and wine anthocyanins for ten red varieties grown in Canada.[162] There was also no correlation between berry size and anthocyanins for five red varieties grown in Spain, or for four red varieties grown in Brazil.[163] The former trial included varieties with 0.7 to 1.9 gram berries, and there was no relation shown between berry size and wine color for any of those varieties.

Table grapes are 3–8 grams or more; winegrapes are typically 1–2 grams. The grapes we use for wine, and that we consider to be the best varieties for wine, are from varieties that do not make big berries in the context of table grapes, although the range in size is still substantial. Casual observations tell us that small-berried Pinot noir, with berry size typically under a gram, is (or was) often lacking in color, and some large-berried red varieties are known for producing strong wine color and flavors, including Zinfandel (2 gm/berry) and Mourvedre (aka Monastrell, 1.75 gm/berry).

In one of the earliest clonal studies to evaluate berry size, a 1988 report showed that although fruit color was inversely related to yield, color was unrelated to berry size.[164] In comparisons of Pinot noir clones in California and Oregon, sugar concentration did not correlate with berry size.[165]

In winegrowing and in research, grapes are usually harvested on the basis of Brix, and this could obscure possible differences among varieties in sugar accumulation and other aspects of ripening due to berry size. Nevertheless, with the available evidence, the only appropriate generalization for genotypes (varieties as well as clones within varieties) is that the concentration of solutes varies considerably, and that this variation is not attributable to differences in berry size governed by the variety or clone. Because the BBB myth is held for both flesh and skin solutes but with different supporting concepts, we will look at flesh solutes and skin solutes independently.

Berry Size within a Variety: Flesh Solutes

By far the predominant solute in the flesh is sugar. Although some early studies reported that the sugar concentration is higher in smaller berries,[166] this seems not to have been borne out by most investigations. Only one of those early studies, conducted with Concord grapes, shows a difference of more than one degree Brix. In simple experiments, entire Cabernet Sauvignon and Chardonnay clusters were dismantled and sampled berry by berry, revealing no relationship of Brix to berry size (the findings for Cabernet are shown in fig. 12A).[167] There is, however, a large range of sugar concentrations among berries on the same cluster—that is, significant variability even within a cluster. Accordingly, analyses of fruit from several Sultana (Thompson Seedless) vineyards found the correlation between Brix and berry size was as often positive (Brix *increasing* with berry size) as it was negative.[168] When Cabernet Sauvignon berries taken at harvest were sorted into size categories, Brix decreased with increasing berry size;[169] yet, when the same type of analysis was conducted with Shiraz, there was no relation between berry size and sugar concentration.[170] These observations indicate that the BBB relationship could be more professed than real.

There is a much clearer relationship between berry size and the total amount of sugar in the berry. Bryan Coombe and colleagues were the first to look directly and carefully at berry size and composition, focusing on sugar accumulation. Coombe posed the question of whether there was a constant amount of sugar per berry or if the amount of sugar increased in proportion with berry size. He measured berry growth and sugar concentration in Dolcetto berries, found reasonably constant sugar concentrations among various berry sizes, and concluded: "The increments in weight of Dry Matter and (Glucose + Fructose) per increment of pericarp (flesh) volume were constant in berries of different size. The increments in Dry Matter, and (Glucose + Fructose) per berry increased with initial berry size, thus discounting a hypothesis of an equal amount of solutes supplied to all berries."[171]

In the other studies, dismantling single clusters or putting berries into size categories showed the same relationship: bigger berries contain a larger solute content (amount of solute per berry). The amount of sugar per berry increased linearly with berry size in the Cabernet size study (r^2 = 0.97), and in another study with Sauvignon blanc (r^2 = 0.96).[172]

While it may seem intuitive that smaller berries are sweeter, the same logic is not generally applied to acidity by the advocates of the BBB myth. Are smaller berries more acidic? This question is not well studied, leading to no consensus among researchers. The concentration of organic acids increases during early berry growth, but decreases during the growth that occurs coincident with ripening. This is another sign that the composition of the berry is dependent on physiological processes other than growth. In the most direct study of this question, Shiraz berries that were separated into size categories did not follow an inverse relation of the concentrations of acids to size.[173]

From the grapevine's perspective, these should not be surprising observations. There is simply more stuff in bigger berries, sometimes a little more one way or the other, but overall berries grown alike reach a point that is generally and similarly ripe, regardless of size. This is coded in the program of berry development; the important issue is to discern what conditions give rise to the "little more one way or the other."

Berry Size within a Variety: Skin Solutes

One arrives at the BBB assumption for skin solutes via geometry. The surface-volume relationship of a sphere = 3/radius of the sphere. This relation says that as the radius of the sphere (berry) increases, the surface-volume ratio decreases. This geometric relationship is used to predict that the ratio of skin to flesh follows the surface-volume ratio,

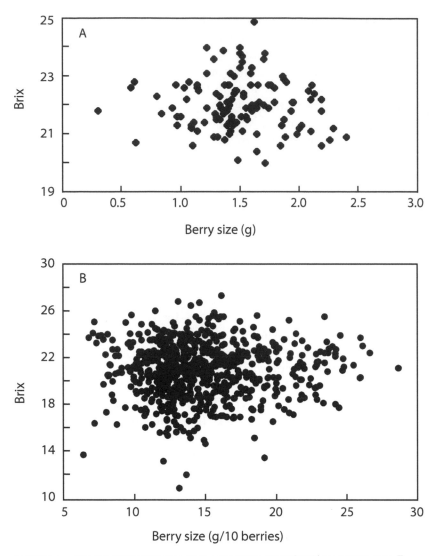

FIGURE 12. THE RELATIONSHIP OF SUGAR CONCENTRATION (BRIX) TO BERRY SIZE. For (A) all berries in a Cabernet Sauvignon cluster (data from Glynn 2003); and (B) combined results from an Oklahoma State University variety trial in Perkins, Oklahoma, with thirteen varieties and several seasons (2003–2010). (Data courtesy of Eric Stafne, Mississippi State University.)

and that as a result, the concentrations of skin solutes are increasingly diluted with increasing berry size. The solutes that accumulate in the skin of the berry, such as tannins and anthocyanins, are inevitably diluted by the flesh (which acts as a solvent) upon crushing. Because reds are fermented on the skins, this hypothesis (that the skin solutes are more diluted in Big Bad Berries) is more important for red varieties. The predicted relationship of skin solute concentration to berry size is shown in the shape of the response curve 3 in figure 6.

The surface-volume relation of a sphere is often repeated in studies in support of the BBB myth:[174]

> Many of the flavor and aroma components are partitioned between the skin and the flesh and the ratio between these two berry parts varies as the berry grows and is a function of their final size.[175]

> As a general rule, wines made from smaller berries will have a higher proportion of skin and seed derived compounds.[176]

> Since the skin to pulp ratio is inversely related to berry size, color extraction would increase during pressing of these grapes.[177]

There are many such examples, all without attribution to data. In this case, I was one of the herd at one point, having claimed that "large diameter fruit would have a greater solvent (flesh cell sap) to solute (skin cell sap) ratio as a result of the lower surface to volume ratio compared to smaller fruit."[178] However, this version of the BBB principle was adopted by others and myself without the benefit of empirical evidence. My comment on color and size was based on the same intuitive assumptions of many predecessors—that the color is in the skin, the skin covers the flesh, the berry is round, and so on. However, the skin is not simply the surface of a sphere; it has mass and volume of its own, and it grows and conducts metabolism. At the time (1980s), there were no data on skin and flesh of the Cabernet franc berries I was studying (or of other berries), so in actual fact, I didn't know what I was talking about.

Fortunately, the skin of grapes can be peeled off and weighed, and its composition can be analyzed. As a berry grows, the skin-to-flesh ratio could theoretically decrease as a fixed amount of skin tissue is increasingly "stretched" over the surface of the berry. The important question of what happens to skin growth as the berry grows, however, had not been addressed at that time. Furthermore, the skin cells in winegrapes look and operate in a different manner than flesh cells (hence they accumulate some different solutes than flesh cells).

My student Gaspar Roby conducted experiments specifically designed to evaluate the role of berry size in the concentration of skin solutes. First, he evaluated the relative amount of skin and flesh tissue to see if the skin decreased compared to flesh with increasing berry size, as predicted by the surface-volume ratio. In the separation of more than 1,000 berries from the same vines, the fresh weight of berries was comprised of approximately 5 percent seed, 15 percent skin, and 80 percent flesh, regardless of total fresh weight or volume of the berry.[179] Several other studies have since shown that the amount of skin increases in proportion, or in approximate proportion, to the size of the berry.[180] Thus, in these winegrapes, the skin grows like the flesh and is not "stretched" around an increasingly large flesh. Although the relative amounts of berry parts can vary depending on variety and environment, the appropriate starting point in thinking about berry size is that, among fruit developing under similar conditions within a vineyard, the growth of skin, seed, and flesh is coordinated.

For berries of different size grown under the same conditions, there may be little variation in the skin-to-flesh ratio, but some vineyard conditions might alter that general relation. For example, in Roby's work, water deficits inhibited growth more in flesh than in skin, effectively increasing the skin-to-flesh ratio, and subsequent studies have found the same effect in other red varieties.[181] Also, when a Barbera grapevine canopy was partially defoliated (presumably exposing fruit to light), berry size increased, but the skin-to-flesh ratio also increased.[182] Thus, although the skin-to-flesh ratio of different-sized berries is often similar within a given environment, environmental conditions can affect the skin-to-flesh ratio as well as overall berry growth. This differential effect on growth within a berry can result in higher concentration of skin solutes such as skin tannins, without affecting how much tannin is synthesized.[183] This result is consistent with the surface-volume version of the BBB, but occurs only under specific circumstances of water deficits after tannin synthesis is completed.

There is also the question of what is in the skin. The surface-volume version of the BBB myth carries the implicit assumption that the amount of skin solute per berry is constant, and that this fixed amount is increasingly diluted by the flesh of ever bigger berries. However, Roby and coauthors analyzed Cabernet Sauvignon berries from vines that were grown under the same conditions, and found that, similar to the amount of skin, the amount of skin tannin and anthocyanin per berry increased linearly with berry size (although the amount of anthocyanin did not keep pace with the size increase). Coombe actually predicted this pro-portionality in 1983, and other studies appear to confirm our observa-tion of proportional increases in solutes per berry.[184] In another study of naturally occurring differences in berry size within a Pinot noir vineyard in Oregon, extracted anthocyanins were unaffected by differences in berry size.[185] With regard to my earlier conclusion that smaller Cabernet franc berries had higher color because of their size, we cannot know how right or wrong that was. The smaller berries were smaller because they developed on vines that experienced water deficits, and water defi-cits themselves have been found consistently to cause increased color synthesis.

VINE SPACING AND BERRY SIZE

The BBB myth is invoked in discussions of vine spacing in the popular press and as part of winery "philosophies" published on various web-sites. For example, in *The Finest Wines of California: A Regional Guide to the Best Producers and Their Wines,* Stephen Brook reports: "In the 1980s, there was a growing realization that wide row spacing, low den-sity, and jungly canopies weren't delivering the desired fruit quality. With European investment came experience of much tighter spacing. Marimar Torres was told by authorities at UC Davis that high density planting was crazy, but she knew she could get better-quality fruit and smaller berries."[186] This is another example of received wisdom that seems to have emerged from undocumented casual observation or imag-ination. The earliest related measurements I could find were in a 1991 South African study on spacing in Pinot noir, where spacing had no sig-nificant effect on berry size.[187] It is feasible that under certain conditions, close spacing would lead to intervine competition for resources that results in smaller berries, as may have been the case for a later South African study on spacing in Pinot noir,[188] but little or no effect is the com-mon observation. Results of three example studies are shown in figure 13. Berry size increased (rather than decreased) with tighter spacing in two cases, and followed the predicted route to smaller berries in the third example of Riesling. More importantly, the change in berry size shown in figure 13A is about 1.5 percent, and in figure 13B is less than 3 per-cent, a difference too small to be worth much effort or cost, even if you could predict in which way berry size was going to respond to spacing. There was also no effect of spacing on berry size in studies on Chancellor in Canada, and in other trials in California: "Vine spacing had no effects

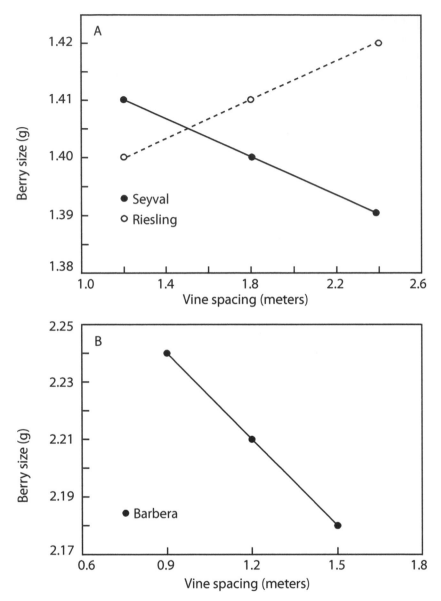

FIGURE 13. RESPONSE OF BERRY SIZE TO VINE SPACING WITHIN THE ROW. (A) Seyval and Riesling grown in Canada (data from Reynolds et al. 2004b and c) and (B) Barbera grown in Italy (data from Bernizzoni et al. 2009).

on grape growth or composition." In Switzerland, "in contrast to large effects on yield (increasing with vine density), row spacing had only a small effect on berry weight."[189]

The BBB concept, with respect to vine spacing, thus lives mostly in the popular press and largely without empirical evidence. Once again, the appropriate starting point for generalizing about fruit growth is different from that indicated by the conventional wisdom. For spacing, one should probably start with the assumption that there is no effect on berry size and evaluate the more specific vineyard conditions (low water or nutrient availability) under which consistent differences in size might be realized.

CHANGING BERRY SIZE IN DIFFERENT WAYS

If the BBB hypothesis portrayed in the wine press is to be believed, fruit or must composition should have lower concentrations of solutes when berry size is larger across varieties, across berry sizes within a variety, and certainly across berry sizes arrived at via cultural practices designed to obtain the desired smaller berries. Berry growth doesn't seem to be governed by vine spacing, but it does respond to crop load, water deficits, temperature, and the berry light environment.

In a recent study comparing kiwi and tomato, fruit growth did not decrease the concentration of total solutes, sugars, or organic acids, and there was no evidence of the concentration of any solute in tomato decreasing during development as a consequence of water import.[190] For most fruit, like tomato and kiwi, big is considered good. The crop is sometimes thinned in tree fruit—but this is usually done in order to *increase* fruit size. Grapevines behave like other crop plants in that fruit size generally diminishes with increased fruit number.[191] While cluster thinning reduces yield, it is not a path to smaller berries. When grapevines are pruned or cluster thinned to reduce yield, berry size generally increases, unless the thinning occurs late enough in the season to prevent a growth response.[192] This sets up a dichotomy for the HYLQ and BBB myths, as it is not possible to get the hypothetical benefits of low yields and small berries if a consequence of low yield is in fact Big Bad Berries.

In a recent study with Cabernet Sauvignon, pruning treatments affected berry size consistently in three consecutive vintages.[193] Wines were made and evaluated each year. The scores of wines were slightly lower for the wines produced from smaller berries. Accordingly, the conclusion of the authors in the Cabernet Sauvignon pruning trial was that "the fact that smaller differences in berry size resulting from vintage differences caused larger differences in wine quality, while larger differences in berry size resulting from treatments caused only small differences in wine quality scores indicates that the relationship among berry size and berry composition and wine quality is neither simple nor direct."[194] We can test the BBB myth in the same way that we tested the HYLQ myth—by changing berry size in different ways and observing whether we get the same responses in fruit composition or not. For comparison, I selected studies in which small berries were achieved three ways: (a) sorting berry sizes from untreated vines (i.e., all grown the same); (b) sorting berries from vines that experienced different water status during berry development; and (c) sorting berries from clusters that were exposed to different light intensities.

First, among the berries grown alike and from the same vineyard block in Gaspar Roby's work with Cabernet Sauvignon, the concentration of anthocyanins decreased significantly as berry size increased (up to 25 percent), in general agreement with the BBB concept. In contrast, the concentration of skin tannin was relatively constant among berry size categories, suggesting that the metabolism that gives rise to the anthocyanins and tannins is important, in addition to any role of berry size.[195] When Roby sorted berries from the same vineyard block into three size categories (diameters of 12.7 mm, 11.4–12.7 mm, and <11.4 mm), made wines, and analyzed each for tannins and anthocyanins, the results were consistent with the berry composition data, in that there were no significant differences in wine tannin concentration among the wines made from the three berry sizes. The wine made from small berries had greater color, and the difference in anthocyanin concentration between the large berry wine and the small berry wine was about 28 percent, similar to the 25 percent difference in the concentration of anthocyanins in fruit among berry sizes.[196] In this and other studies, both fruit and wine composition were much less sensitive to fruit size than what is predicted from differences in surface to volume of the berries.[197]

In other studies in which wines were made, an Australian trial on Shiraz reported berry size had no significant effect on juice potassium (another skin solute) or the organic acid tartrate, and Brix and pH showed no consistent trend with berry size. Wine characteristics, including color parameters, were similar from both small and large berry lots.[198] In another study, Rob Walker and colleagues in South Australia separated Shiraz berries produced from the same viticulture treatment into "small" and "large" groups to make separate wines; the two wines showed no difference in quality scores.[199] Other experiments on berry size in red winegrapes and wines found little effect on the wines.[200] Within a red variety and depending on the means of changing berry size, there is sometimes a BBB-consistent change in composition, more consistently for color, less so for tannins, and rarely for potassium or acidity.

The second method of comparison is accomplished by evaluating the concentration of solute in berries of the same size that developed under well-watered versus water-stressed conditions. It is well established that water deficits inhibit berry growth, and, when the timing is correct, increase the concentrations of skin solutes such as anthocyanins and tannins. By sorting fruit into sizes after exposing vines to different irrigation regimes, it is possible to evaluate whether or not the increased concentrations arise simply from inhibited berry growth.

When comparing berries of the same size, any difference in concentration cannot be due to growth, dilution, or size, and should be attributed to other environmental effects on berry-ripening metabolism. For berries of the same size, skin tannin concentration increased 22–28 percent and anthocyanin concentration increased 15–33 percent in berries grown under water deficits, compared to those grown on well-watered control vines.[201] The results showed that there are size effects on color, but also large effects of vine water status on color and tannins that arise independent of any effects of fruit size. Furthermore, the effect of vine water status on the concentration of skin tannin and anthocyanins was greater than the effect of fruit size on those same variables. Unless one assumes beforehand that the growth of the berry is the only means by which the plant responds to the environment, these results should not be surprising.

A third means of manipulating berry size is via the cluster microclimate. Mark Kliewer and Nick Dokoozlian grew Cabernet Sauvignon and Pinot noir in Davis, California, with clusters exposed to various light intensities, from darkness up to about 20 percent of full sunlight. The different light intensities produced berries of different sizes; berries grew more in increasing light. In Dokoozlian's experiments, Brix, anthocyanins, and total phenolics were measured at harvest.[202] When these aspects of fruit composition are plotted against their respective berry sizes, the results are striking. The concentrations of all three solutes increased dramatically with increasing berry size. For both flesh and skin solutes in both Cabernet Sauvignon (shown in fig. 14) and Pinot noir, the increase was approximately linear over the range of low

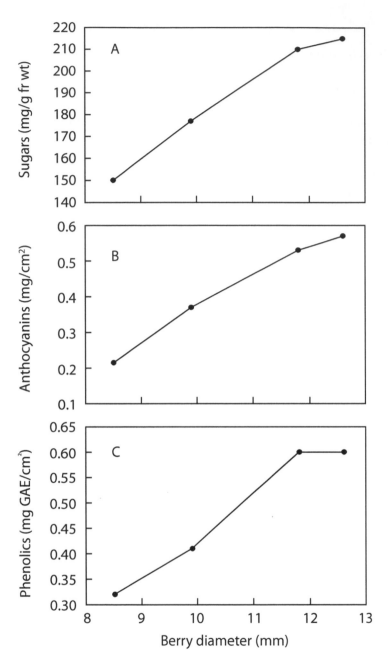

FIGURE 14. DEPENDENCE OF SOLUTE CONCENTRATIONS ON BERRY SIZE FOR
CABERNET SAUVIGNON BERRIES GROWN IN DIFFERENT LIGHT ENVIRONMENTS.
Skin solute concentrations are given as mass of the solute per cm^2 of skin
area; phenolics concentration is given in gallic acid equivalents (GAE) due to
the particular assay used to measure phenolics. Wines were not made in this
study. (Data from Dokoozlian 1990.)

light intensities studied. The results of the shading experiments could hardly be more different from the predictions of the BBB myth.

Although Dokoozlian and colleagues reported some similar results in a later field trial, the light response of growth and composition is not always the same. Berry size is generally *reduced* by shading, although there are counterexamples.[203] In another study, when Cabernet Sauvignon grapevines were exposed to from 2 to 100 percent sunlight, there was no effect on berry size, but a large effect on anthocyanins was observed.[204]

One would think that these and other possibilities have been investigated and assimilated into the zeitgeist that is the BBB paradigm, but in fact, the relationship of berry size to composition has hardly been investigated. Nevertheless, it would be a mistake to think that the data in figure 14 are the odd exception to BBB. A few additional examples: When Shiraz was grown on two rootstocks, berries were slightly bigger on one rootstock; however, the larger fruit had higher concentrations of sugars and anthocyanins, not lower.[205] When berry size was decreased by growth in high temperatures, the concentration of anthocyanins (berry color) was greatly reduced, rather than increased as predicted by BBB.[206] In another contrast to the BBB myth, when table grapes are increased in size by applications of the growth regulator gibberellin, sugar concentration is usually unchanged or increased.[207] In two studies that reduced functional leaf area dramatically by pulling off leaves or shading them, berry growth was inhibited and the smaller berries had lower (not higher) sugar concentrations in one study,[208] and there was no relation of size to Brix when berry size was reduced 20 percent in the other one.[209]

The surface-volume version of BBB is usually employed in conjunction with red varieties. However, the skin is also where MIBP resides in Cabernet Sauvignon, Sauvignon blanc, and related varieties. The strong veggy character of MIBP is in disfavor in most reds, and this can present a dichotomy for winemakers' goals via BBB thinking, in that small berries should have higher MIBP concentrations. Only color and tannins have been given significant attention in research related to berry size, but one study reported on MIBP concentrations in Sauvignon blanc berries, where the MIBP concentration was more (rather than less) in the larger of two berry size classes.[210] While this may be good news for those hoping to take advantage of small berries in terms of other desirable skin solutes, if confirmed, it is another problem for the BBB concept, because the concentration of the skin solute MIBP did not follow the BBB prediction. The widely varying relations of fruit composition to size suggest that the conditions giving rise to berry size (not just the berry size per se) are fundamentally important to the final fruit composition.[211]

CONCLUSIONS

The HYLQ and BBB ideas are attractive in their intuitive and simple nature, and proponents tend to make sweeping declarative statements that embrace both myths as fundamental truths about grapevines. However, the assertions about grapevine biology from the wine press are seldom accompanied by supporting and verifiable evidence. Although the direction of research has been largely to test the various practices already in place, rather than to test the HYLQ or BBB concept directly, there is sufficient evidence to draw some important conclusions.

If these concepts are indeed key aspects of winegrape quality, then the various means of attaining low yield and small berries should bring the desired results. I selected three common means of changing yield for comparison. When the destination (goal) of lower yield was attained by pruning, cluster thinning, and vine water status, the increase in desirable flavor intensity predicted by HYLQ was not observed consistently (fig. 15). Wine sensory attributes did not differ by just degree or intensity; they differed by the *direction* in which the flavor and aromas changed. Wines became increasingly veggy with pruning-controlled yield, but became less veggy with water deficit–controlled yield. Furthermore, the intensity of vegginess and fruity aromas and flavors moved in opposite directions within the same changes in yield in the pruning trial. Results were similar for the intensity of vegginess and fruity aromas in the irrigation trial, but in the reverse direction, as in the pruning trial. These relationships were not observed among the wines in the cluster-thinning trial, where changing yield in that manner had little impact. In fact, the changes in wine sensory attributes were not great overall, but nevertheless, three journeys to lower yield produced three different flavor responses in the resultant wines.

Examples of qualitatively different responses in fruit composition are also found when the destination of small berries is reached by sorting berries grown under the same conditions, by vine water deficits, or by cluster shading (fig. 15). Small berries from a cohort grown alike (no treatment) had more color and little or no change in tannins. Small

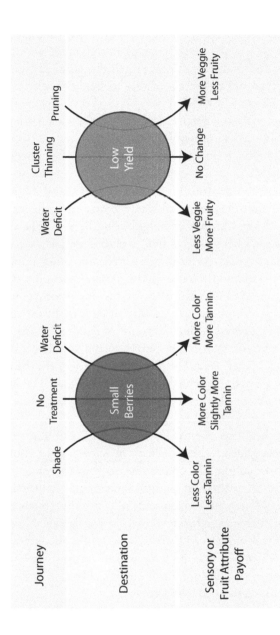

FIGURE 15. IT'S NOT THE DESTINATION, BUT THE JOURNEY THAT AFFECTS RESULTANT FRUIT AND WINE SENSORY
ATTRIBUTES. This diagram summarizes experimental work comparing the effects of using various means to change berry
size (*left*) or vine yield (*right*). When the destination of small berries was arrived at via water deficits, color and tannin of
the berries increased, but when the destination of small berries were arrived at by shading, color and tannins of the berries decreased.
When the destination of low yield was arrived at via pruning, wines were more veggy and less fruity, but when low yield
was arrived at by water deficits, wines were less veggy and more fruity, and when low yields were arrived at by cluster
thinning at veraison, there was little effect on the sensory attributes of the resultant wines.

berries obtained by water deficits had increased color and tannins (qual-itatively consistent with BBB);[212] however, small berries arrived at by shading had the opposite responses, with less color and tannins than the big berries (in direct contrast to the BBB prediction). Changes in fruit composition were large when berry size was arrived at via water deficits or shading, less so for fruit from vines treated the same. As with yield, there were three responses to three journeys to smaller berries.

The specific journeys to yield and berry size affect berry metabolism and flavor development, and often more so than the yield or berry size itself. There is no biological principle that tells us a high concentration of flavor solutes is linked to low yield potential. (And the concentration of solutes in plant parts is not generally inversely related to size or yield.) Fruit composition depends on fruit metabolism and its responses to an environment that include much more than simply the size of the individual berry, or the number of neighboring berries. The environ-mental journey of each berry affects fruit metabolism and flavor devel-opment independently of yield and berry size, and even impacts the relative amount of skin and flesh in individual berries (of any size).

Interestingly, the HYLQ and BBB myths incorporate mutually opposed assumptions about the availability of flavors. In order for the HYLQ idea to work, there is a fixed amount of flavor per vine, leaving a decreasing amount of flavor per berry as the number of berries increases to make a higher yield. For the BBB idea to work, the amount of flavor per berry stays fixed, while the berry gets bigger. The only way that HYLQ and BBB could work together is for all changes in yield to be due to berry size, and this is rarely the case.

There are, however, important ways in which HYLQ holds true and the BBB may reflect reality. The reduction in wine supply afforded by restricted grape production can contribute to upward price pressure (which is one means of evaluating quality); and greater fruit maturity is possible at low yield when an environment restricts how much crop can be ripened.[213] The former situation involves a conflation of price and quality that will not be addressed further here. As for the latter, HYLQ generally applies whenever it is difficult to get grapes ripe before the growing season comes to a close. This could arise because of limitations by soil, climate, or genetics (i.e., the variety).

These situations may lead to the HYLQ scenario depicted in curve 2 of figure 6, in which quality (ripeness) drops off at high yields. In these ways, the intuitive HYLQ notion may have been *a part* of the actual story, but it turned into a widely held belief before being tested for its robustness. Claims of high-yield sensitivity in some important winegrape varieties have not been borne out in studies measuring fruit and wines from vines with greatly differing yields.

It is a similar scenario with berry size: under conditions that are not yet resolved, smaller berries have higher concentrations of some skin solutes, especially color. With today's optical fruit sorters that can identify each berry color, shape, and size, it will be interesting to see if small berry wines achieve the desired effects. I think that both yield and berry size may respond in general as curve 2, but don't know if the generalization is consistent enough to be useful.

HYLQ and BBB have been sustained as popular paradigms in the face of much contradictory evidence. Yet, an assumption of no relation between yield and solute concentrations is the commonsense starting point for a discussion of yield and fruit quality, certainly for cross-variety comparisons, but also within a variety. For fruit at similar maturity, the empirical evidence argues strongly that factors other than yield (such as water availability) are important in determining fruit composition. While these factors may sometimes correlate with yield, they are at least as likely, if not more likely, to have a causal effect on final grape and wine composition than yield or berry size, per se. There is still not enough straightforward investigation, completed over a wide-enough range of yields, varieties, and environments, to inform us quantitatively in a way that validates generalizations beyond this point.

Many viticulturists feel that the yield data have not borne out the HYLQ myth, because of its oversimplification, which does not recognize that fruit ripening is dependent on the source leaves. From that recognition, the concept of "vine balance" has emerged to account for yield-quality relations by employing a more plant-based concept. In the next chapter, we will look at this metric of vine balance, and investigate whether or not it has contributed to our understanding of fine winegrowing.

CHAPTER 2

Vine Balance Is the Key to Fine Winegrapes

Concepts which have proved useful for ordering things easily assume so great an authority over us that we forget their terrestrial origin and accept them as unalterable facts. They then become labeled as conceptual necessities, a priori situations, etc. It is therefore not just an idle game to exercise our ability to analyze familiar concepts, and to demonstrate the conditions on which their justification and usefulness depend, and the way in which these developed.

—Albert Einstein

Viticulturists generally line up behind the concept of "balanced vines" as the key to winegrape quality. "Vine balance" refers to the putative balance between vegetative growth (leaves, stems, roots) and reproductive growth (flowers and berries) that is said to create the best winegrapes. Winemakers and wine writers also refer to the balanced vine; according to the website of Stonum Vineyards (headquartered in Lodi, California), "Vine balance equates to flavorful wines of quality and distinction"; but the concept's origins rest with the viticulture academics. This balance, attained theoretically by judicious vineyard management, is believed to be a fundamental principle of fine winegrowing. On the one hand, balance is intuitively good. On the other hand, what does vine balance truly mean, and how does one know when it has been achieved?

In everyday language, "balance" is usually employed in the sense of a state of equilibrium, as with two sides of a fulcrum. In viticulture, vegetative and reproductive growth are the two sides that must be balanced, but the real concern in terms of final quality is with the grape composition. Another common use of "balance" is in reference to an aesthetically pleasing integration of elements. Vineyards are typically appealing to the eye, attracting painters and photographers in addition to wine enthusiasts. Does a more visually pleasing vineyard produce a more flavorful wine? One might be inclined to think so, and this (potentially subconscious) thought may be involved in the analysis of whether or not a vineyard is in balance. This may seem improbable, but lest you quickly dismiss this idea, note that many a vineyard manager or consultant will eyeball a vineyard to assess whether it is in balance. A third use of "balance" refers to equilibrium in the sense of long-term stability. Grapevines are perennials with variable budbreak in the spring (some buds come out of dormancy and grow while other buds remain dormant) and indeterminate flowering (the number of flowers and timing of flower development are variable). The extent of both processes varies from year to year, a result of weather and vineyard management practices. Although not part of the contemporary sense of vine balance, the long-term stability of budbreak and fruitfulness also deserves consideration as a meaningful sense of vine balance.

Several leading viticulturists have been moved to write in an effort to define vine balance.[1] Each of their essays starts from the premise that a balance of vegetative and reproductive growth exists, and that attaining this balance is a necessary condition for fruit and wine quality. Yet each author also wrestles with the meaning of a balanced vine and how to know when this balance is achieved. Together, the essays contain no less than six different definitions or metrics of vine balance. The various metrics (such as yield/pruning weight, yield/leaf area) are each believed to evaluate the same thing: the balance, or lack thereof, between vegetative and reproductive growth.

The vine balance concept and its utility in viticulture will be investigated in this chapter, but first I want to acknowledge that plant growth is clearly constrained by other well-established balances. Two clear examples of balance present in plants are root-shoot growth and the relationship between fruit size and number.

ESTABLISHED BALANCING ACTS IN PLANTS

Root-Shoot Balance

There is a necessary balance between shoot and root growth—each provides what the other needs in order to sustain growth: roots deliver water and mineral nutrients; shoots (leaves) supply photosynthate. When conditions such as shade limit the shoot function (photosynthesis), shoot and leaf growth is enhanced at the expense of root growth, and when nutrient availability is low, root growth is enhanced at the expense of shoot growth (fig. 16A).[2] Another example of root-shoot balance can be seen when cutting roots temporarily leads to root growth, at the expense of shoot growth, until the balance is recovered—as shown in figure 16B. This balancing reaction has been observed in the vineyard, when roots, shoots, or crop is removed, resulting in a linear relationship between shoot and root growth.[3] The vine balance idea in viticulture, however, addresses only above-ground growth and ignores plant roots.

Source-Sink Relations of Fruit

One way that plant biologists envision plants is as a system of "sources" and "sinks," in which the classification of a plant part is based on whether photosynthate (or other plant nutrient) is distributed from (source) or delivered to (sink) an organ or tissue. Leaves are primary sources and roots, stems, and berries are sinks. The source-sink analysis is commonly applied to whole plant growth—that is, roots, shoots, and other organs, including trunks and fruit in grapevines.

Crop plants are evaluated as being source limited or sink limited. When source limited, insufficient photosynthate is delivered from leaves to all the crop sinks, preventing them from growing to their full potential. When sink limited, not enough sinks are present to take advantage of the available photosynthate. When crops are source limited, increasing the number of fruit results in smaller fruit. When sink limited, increasing the number of fruit does not impact fruit size—there are just more, big fruit. In general, yield increases with the number of fruit per plant when sink limited, and does not increase when source limited. The middle, where increasing fruit number still increases yield, but fruit size begins to decrease, could be considered a balance point between source and sink limitation. Also somewhere in the middle is an economic optimum (which could also be considered a balance point) for returns on the fruit yield for fruit that are judged on size. These growth relations are intensively studied in some crops, particularly tree fruit, but not so much in winegrowing, where yield and fruit size are considered less important than aspects of ripening.

The concept of vine balance is associated with these well-established balance phenomena in crop plants, but it is different in a very important way in that it claims that the *quality* of the fruit depends on a certain

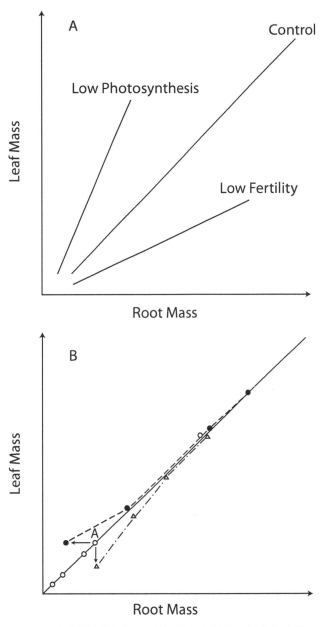

FIGURE 16. THERE IS A GENERAL BALANCE, OR HOMEOSTASIS IN
FUNCTIONS, BETWEEN ROOT AND SHOOT GROWTH IN PLANTS. (A)
When low fertility limits root function (Low Fertility curve), shoot
growth is reduced relative to root growth. When leaf function is
reduced, e.g., by shade (Low Photosynthesis curve), shoot growth
is promoted relative to root growth. (B) When shoot or root
growth is pruned, subsequent growth moves in a direction that
seeks the original balance. (Figures redrawn from Rendig and
Taylor 1989.)

balance, as opposed to the previous concepts, where the outcomes are simply a balance between the number of roots and shoots or the size and number of fruit. Vine balance recognizes that there is a requirement for some amount of leaves (leaf area) to conduct sufficient photosynthesis to grow and ripen fruit, but goes beyond that to declare that fruit quality depends on the relationship between mass of leaves and mass of fruit.

The story of vine balance involves a complicated history, in part because many viticulturists take it to be an a priori fact, yet its specific meaning has wavered and remains elusive. Vine balance has referred to both vine fruiting and to grape quality, two distinctly different aspects of the grapevine. In addition, contemporary viticulturists recently developed a history of vine balance that is poorly documented. In the following sections, that history of vine balance is critiqued, and a new version based on my research is presented.

THE CURIOUS TALE OF THE RAVAZ INDEX

Australian viticulturist Peter Dry and colleagues, along with most viticulture authors, give famous viticulturist Louis Ravaz (1863–1937) credit for introducing the concept of vine balance into viticulture; however, the path to this attribution is odd. Ravaz was a director of the viticulture school, now the Institute for Higher Education in Vine and Wine at Montpellier SupAgro, in Montpellier, France.[4] Ravaz made many scientific contributions around the turn of the twentieth century. He played a prominent role in the late nineteenth-century battles over whether to deal with the devastating root louse phylloxera by using hybrids (crosses of traditional *Vitis vinifera* with resistant species) or grafted vines (*V. vinifera* varieties grafted onto rootstocks from resistant species).[5] Ravaz and his colleague Pierre Viala discovered the causal organism behind the fungal disease "black rot" on grapes,[6] and Ravaz published several important articles and books from 1887 until 1937, including *American Vines: Their Adaptation, Culture, Grafting, and Propagation*. Ravaz's work was cited in research papers into the 1940s.

Fifty years later, viticulturists began citing Ravaz once again. He is given credit for introducing both the vine balance concept[7] and the first and primary metric of vine balance in winegrape production: a ratio of the weights of crop yield and subsequent prunings, expressed as yield:pruning weight (Y:PW) and aptly named the Ravaz Index. This is curious, because not only was Ravaz's work essentially un-cited for a half century prior, but he was also not cited for vine balance or his Y:PW ratio in scientific papers *at all* previously.

The first mention of the Ravaz Index in the peer-reviewed literature was evidently by leading Italian viticulturist Attilio Scienza and colleagues in proceedings papers from "Strategies to Optimize Wine Grape Quality," an International Society for Horticultural Science workshop held in Conegliano, Italy, in July 1995.[8] Almost immediately, several other groups employed the Ravaz Index with no citation, as though the term was part of the common understanding in viticulture.[9] Similarly, Michigan State University professor Stan Howell wrote of vine balance in 1998 without mentioning Ravaz, although his 2001 review article included the Ravaz Index as a key word.[10] It is easy to see that references to both the Ravaz Index (fig. 17A) and Louis Ravaz as the source of the vine balance concept (fig. 17B) took off after 1996. Yet no viticulturist has commented in the literature on what happened to bring the Ravaz Index about, let alone reviewed the evidence that led to its adoption.

Scienza and Howell both report that Ravaz established a close relationship between vine leaf area and vine pruning weight, so that pruning weight could be used as a proxy for leaf area. They cite different Ravaz studies, but remarkably neither cited paper establishes the relationship as claimed.[11] There are no leaf area measurements in either paper, although in a footnote, Ravaz does remark that there *should* be a close relationship. Another attribution of the Ravaz Index is an obscure research note from 1903 on *brunissure* (brown leaf or vine browning, a condition referred to as a "disease" in the early 1900s) cited in a review article in 2000.[12] There are no data for Y:PW, and no balance mentioned in that paper, yet more than a dozen subsequent papers cite Ravaz's 1903 publication when referring to the Ravaz Index.

Nevertheless, Ravaz *did* introduce the Y:PW ratio as a metric in winegrowing, using it for the first time in a 1902 report on brunissure.[13] At the time, he was not trying to find a vineyard metric for how to grow great wine; he was studying the leaf browning problem in young vines that were three or four years old. He selected two to six vines each from about twenty varieties and labeled them sick or healthy, based on his sense of the extent of leaf browning. (That he considered shoots with up to ten brown leaves to be "healthy" implies that most shoots and perhaps all of the vines in his study were afflicted to some extent.) Ravaz then collected yield and pruning weight data for each vine, and reported yield and the Y:PW ratios in a table. It was obvious to him that the sick vines had the higher crop, and from that Ravaz concluded that the

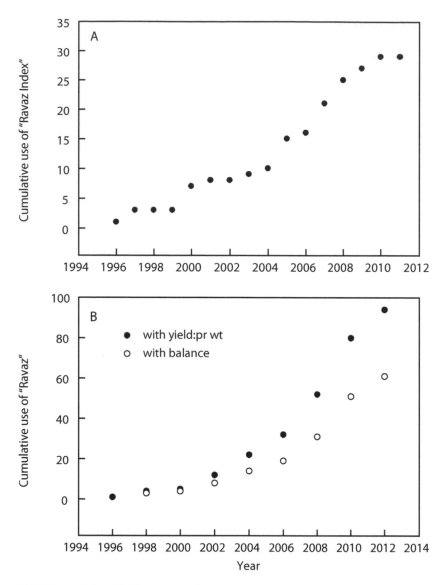

FIGURE 17. THE RECENT DEVELOPMENT OF SCIENTIFIC REFERENCES TO THE TURN-OF-THE-TWENTIETH-CENTURY WORK OF LOUIS RAVAZ IN THE CONTEXT OF "VINE BALANCE." (A) The cumulative number of papers using the "Ravaz Index" originated in 1996 and increased steadily thereafter. (B) The cumulative appearance of "Ravaz" together with "yield:pruning weight ratio," or with "balance." (Data from CAB abstracts.)

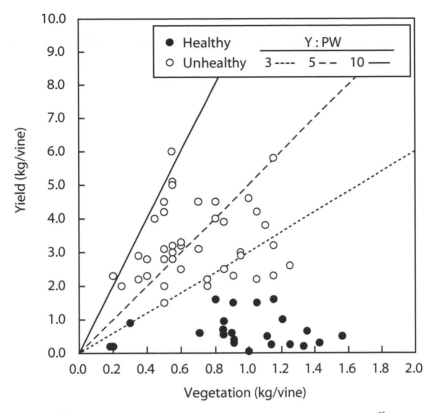

FIGURE 18. PLOT OF ORIGINAL RAVAZ DATA ON YIELD AND PRUNING WEIGHT. Shown as yield at various pruning weights (vegetation) for healthy vines (open symbols) and sick vines (closed symbols). Lines are for Y:PW values of 3 (from Ravaz), and for 5 and 10 (from current dogma). (Data are from Ravaz's 1902 paper on brunissure.)

higher crop caused browning, claiming, "I feel no need to give other examples." There were no treatments or experiments included in the report, just observations on vines with varying amounts of leaf browning, and an inferred relationship between yield and brunissure.

Ravaz suggested that a Y:PW ratio of three may be the maximum that could be sustained without danger of serious leaf browning. Ravaz's suggested cutoff is illustrated by the dotted line in figure 18 (when plotting yield vs. pruning weight or "vegetation," the slope of a line is a Y:PW ratio). This is the origin of the Y:PW ratio: a metric proposed in 1902 to prevent browning and keep leaves on young vines looking healthy.

Two years later, Ravaz published a booklet on brunissure in which he reiterated that the Y:PW ratio should be less than three for young vines, but added that it could be up to nine in older vines.[14] In fact, Ravaz concluded that browning was a disease of young vines caused by over-production, and that, by his own account, vines will grow out of it: "It is therefore easy to avoid this affliction which, moreover, will become less and less frequent as the vines become older or more developed: browning is a malady of young vines."[15] It turned out that in the man-agement of a mature vineyard, the leaf browning problem seemed to just go away. Problem solved?

Whether the true solution to brunissure had been found or not, Ravaz became convinced that high yield was a central problem in viti-culture, and the way to identify the hazard was via Y:PW. In his 1906 report on "overproduction," Ravaz stated:

> It's a question that I've already studied as regards the browning of the vine. I showed that overproduction always brings a weakening, whose only limit is the death of the plant.
>
> As such, the cases of fading indicated in some parcels of vines in Tunisia are not due to a malady; we must consider them accidents whose cause is per-fectly well known. This cause is overproduction, or the disproportion that exists between the capability of the strain (variety) and the number of grapes that it yields.[16]

Overproduction was thought to be so powerful that it could kill vines. In California, a disease swept through vineyards in the 1880s and 1890s. The disease, which devastated vineyards in the Los Angeles and Santa Ana Valley areas and was at the time called Anaheim disease or California vine disease, had no known cause. Ravaz's colleague Pierre Viala independently claimed in 1892 that brunissure was caused by the fungal pathogen *Plasmodiophora vitis*, and California vine visease was caused by *Plasmodiophora californica*. Soon after, Ravaz began to argue that brunissure and California vine disease were one and the same, and that both were a consequence of overproduction.[17] This brought him into conflict with his colleague Viala. Viala's fungal species was eventually proven to be a "phantom species"—one that was not really there at all.[18] Of course, a phantom species could not be the cause of California vine disease or brunissure. Ravaz was also wrong in his belief that brunissure and California vine disease were the same, and wrong again in claiming that the disease was due to overproduction and thus could be cured by a lower Y:PW. Much later in the 1970s, the disease was renamed Pierce's disease, and it was found to be caused by the xylem-dwelling bacterium *Xylella fastidiosa*.

In the 1950s, plant pathologists at UC Davis briefly reviewed Ravaz's work and suggested that he was studying vines with leafroll virus when making his Y:PW analyses.[19] The leafroll virus produces symptoms that led to the disease being called "red leaf," and the thought was that "brown leaf" and "red leaf" could both be names for leafroll virus disease. The leafroll virus hypothesis is not consistent, however, with Ravaz's observation that vines grow out of brunissure, because leafroll usually gets worse with time, rather than better. Nevertheless, viruses and other diseases were probably common during the time of Ravaz's work, given that there were no programs for disease-free planting stock and there were limited control measures in place. Diseased vines have impaired leaf function and fruit ripening, and if vines at that time were often diseased,[20] getting fruit to adequate sugar was necessarily an overriding concern due to disease, rather than an inherent physiological constraint.

The grapevines during Ravaz's research likely had other problems as well. Later research, some of which was conducted by Ravaz himself, showed that brunissure was caused by potassium deficiency (which is corrected with potassium, rather than pruning).[21] Accelerated leaf senescence is characteristic of mineral deficiencies and often results in a reddening or browning of leaf tissue, particularly in red grapevine varieties. Ravaz and others reported that applications of potassium ameliorated the leaf symptoms. Accordingly, references to brunissure in French peaked in 1910 and in English in 1920,[22] and instances of "brunissure" in viticulture faded almost completely away, except as an occasional reference as a symptom of or synonym for potassium deficiency.[23]

Ravaz and Wine Quality

Ravaz had surprisingly[24] little to say about wine based on his vineyard research on Y:PW, but he did make some wines and collected limited data. Sugar (or the resultant alcohol in wine) was the only aspect of quality that Ravaz addressed directly. Although he sometimes reported on acidity, there is no indication that Ravaz used anything other than must density (determined by sugar concentration) or wine alcohol to infer quality. In the earlier discussion of the High Yield Low Quality concept, we noted that higher levels of sugar occurring with lower yield is likely the most consistent observation in experimental viticulture.

Practical experience must have established this early on. The challenges of keeping vines alive and obtaining ripe and sound fruit were at the top of growers' concerns, especially given that Ravaz's work took place prior to so many advances in the management of grapevine pests, diseases, and nutrition. Accordingly, Ravaz equates the concentration of sugar in the must or alcohol in the wine with wine quality (his original graph labeled the alcohol concentration data as "Quality," as shown in fig. 19B).

Ravaz's data show that quality simply decreases as F/V (i.e., yield:pruning weight) increases (fig. 19A); there is no optimum or balance point. Ravaz used his data shown in figure 19A to draw the curve in figure 19B, and may have thought he had found a sweet spot in Y:PW that would be a quality target by emphasizing the wiggle he put in the curve. Yet, his suggested Y:PW at that wiggle would produce wines with only about 8 percent alcohol! The last and most clear reference Ravaz made to balance that I could find was in one of his last papers, a 1935 review, "Factors of Quality (in Wine) and Their Relation to Agricultural Practice," in which he said that there was a pruning balance that produced the best wine, but he did not use Y:PW to define it, nor did he refer to any data to support his contention.[25]

Improved viticulture practices (and global warming) have alleviated much of the concern about reaching a good sugar concentration, although this issue is still faced in cooler winegrowing regions today. It is the rare winemaker who says that he or she harvests on the basis of sugar concentration alone, yet Ravaz's work used only sugar for quality, and repeatedly indicated that he simply thought less yield was better, period. There is no mention of too little fruit in Ravaz's writing; generally he writes that only good things can come from growing less crop. The incorporation of Ravaz into the literature as the source of the vine balance concept in winegrowing was remarkably fast, and this may have led to some mistakes. None of the contemporary studies citing Ravaz as the origin of vine balance (as expressed in Y:PW) mention the context of his work to avoid leaf browning symptoms or vine exhaustion, the discovery of potassium correction for leaf browning, the little exploration that he did on wine quality, or his mistaken diagnosis of Pierce's disease as a consequence of unbalanced vines.

Furthermore, the idea of balancing vegetative and reproductive growth was around before Ravaz's limited work on quality, and was expressed explicitly as early as 1888 by George Husmann in a report on viticulture to California state authorities:

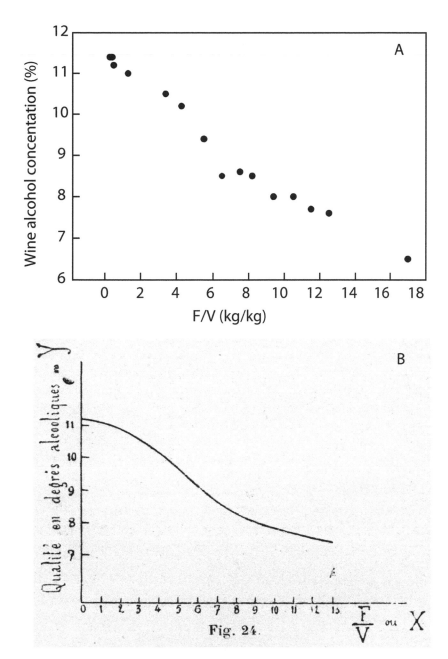

FIGURE 19. THE RELATIONSHIP BETWEEN WINE ALCOHOL CONCENTRATION AND Y:PW (F/V IN RAVAZ'S NOTATION). (A) Plot of Ravaz data, originally presented in a table in Ravaz 1904a, 115. (B) Copy of the figure as it appears in Ravaz 1904a, 116, which Ravaz evidently created from the data plotted in A. (Figures A and B are taken from Ravaz's 1904 book on brunissure, *La brunissure de la vigne: Cause, conséquences, traitement.*)

In my opinion the greatest perfection of the grape depends upon having just as much to bear each season as it can ripen in perfection. If we overload it, inferior insipid fruit will be the result and a feeble growth of wood which will also not ripen fully. If on the contrary we prune too short a rank succulent growth, black knot, coulure [dropping of flowers/berries], etc. will be the result, and the fruit will also suffer accordingly. On this nice balancing of the powers of the vine more of the success of the vintner depends than many are aware of.[26]

It is therefore problematic to attribute the vine balance concept to Ravaz without further evidence to support the assertion that a century ago he elaborated one of today's most cherished viticulture principles.[27] It would be far better to give Ravaz credit for having recognized the following insight:

It is accepted generally that "quality" and "quantity" are incompatible. One excludes the other. They are therefore in a relationship which, as formulated, is very simple. Perhaps, however, this simplicity is more apparent than real and that's what we will find. . . . The quality of the grapes or wine depends on the absence or presence and proportion of some bodies, some of which are very poorly known, while on the other: sugar, alcohol, tannin acid, etc., we have knowledge of a little more complete.[28]

Ravaz predicted the received knowledge that yield and quality are mutually exclusive would be found incorrect; he recognized that factors other than sugar are important to wine quality, and that the existing knowledge of those factors was quite poor.

ORIGINS OF VINE BALANCE IN THE EXHAUSTION CONCEPT

In an excellent short review on vine balance, Peter Dry and colleagues state, "For as long as grapes have been grown, it has been known that the best wines come from those vineyards where vegetative growth and crop yield are in balance."[29] They don't offer specific support for that contention, perhaps because many viticulturists feel that vine balance exists as an a priori fact. However, vine balance may have evolved from a demonstrably ancient and related concept—exhaustion. Exhaustion was the idea that organisms can injure themselves by excessive reproduction. It was reported by Aristotle for fruit trees:

Most trees, if they bear too much fruit, wither away after the crop when nutriment is not reserved for themselves.

And similarly for chickens:

> And some fowls after laying too much, so as even to lay two eggs in one day, have died after this. For both the birds and the plants become *exhausted,* and this condition is an excess of secretion of residual matter.[30]

We don't know what Aristotle saw. Perhaps some fruit trees and chickens met untimely ends, but the cause he attributes seems unlikely. Ancient reasoning relied in part on analogies, but analogies between animals and plants can only go so far. Plants don't study or run laps, and they do not become mentally or physically exhausted like humans do. A logical extension of Aristotle's exhaustion idea is that high producers are selected against, so that today we would be stuck with the survivors of this tragedy repeated over and over—in essence, the good producers got rubbed out, and we are living off of the low-yielding chickens, trees, and vines whose feeble production allowed them to escape the withering away caused by exhaustion. Yet productivity of today's tree fruit, chickens, and grapevines is greater, not less, than it was in the case of their forebears.[31]

The exhaustion concept was generally accepted in vineyards at the turn of the twentieth century.[32] Frederic Bioletti (1865–1939) was a contemporary of Ravaz, and the University of California's first professor of viticulture. Following the sense of predecessors in early California viticulture, Bioletti warned in the late 1800s that in some production systems there is a tendency toward overproduction and consequent exhaustion of the vine; in 1913 Bioletti wrote explicitly that managing vines to promote fruitfulness may lead to "premature exhaustion."[33] Similarly, Ravaz refers to the "fading" of vineyard blocks that he attributes to putting on more crop than the variety can sustain.[34]

Ravaz, Bioletti, and others referred to a hastening death, as though allowing vines to bear fruit accelerated an aging process, but this again may be too literal an analogy. To understand why this line of reasoning is problematic, let's briefly review a few fundamentals in plant development and aging. An annual plant, such as corn, must flower and make seed in the current season in order to sustain another generation as a new plant. As such, the whole corn plant senesces and dies in conjunction with the completion of seed development. Whether we consider flowering as detrimentally causing senescence, or exhaustion, is immaterial to the biological fact of the growth habit of the species.

For annual plants, there is no alternative to flowering and death. In contrast, as an iteroparous perennial, the grapevine makes fruit and

TABLE 4 EXAMPLES OF THE LONG LIFE SPANS IN SOME PERENNIAL PLANTS

Perennial plant	Maximal life span
European white birch (*Betula verucosa*)	120 years
English ivy (*Hedera helix*)	200 years
Apple (*Pyrus malus*)	200 years
Grapevine (*Vitis vinifera*)*	250–400 years?
Scots pine (*Pinus silvestris*)	500 years
Olive (*Olea europaea*)	700 years
Giant sequoia (*Sequoia gigantea*)	3,200 years
Bristlecone pine (*Pinus aristata*)	4,600 years

*Grapevine life span added by author based on ages claimed for various old vines around the world.

seed many times from many plant parts. In temperate climates, the grapevine goes dormant each fall and begins growth anew each spring. "Individual modules (root or shoot branches, inflorescences, leaves) are dispensable for the survival of the organism, and their function can be replaced by tissues differentiated from new cells that are made each season. In this regard, lifespan is largely defined by the indeterminate growth (making flowers or not) of vegetative meristems."[35] The grapevine can flower, make fruit and seed, or not, and still live on into the next season, because it also develops buds that serve as an alternative overwintering form. Thus, the grapevine does not senesce like corn, but parts of vines, such as leaves and fruit, do senesce each year.

Grapevines do not live forever. It is *possible* that fruiting accelerates vine aging, but we don't know what causes perennials to age and die in the first place.[36] From research in other plants, we find that "an age-dependent decrease in plant fitness has been well documented in some perennials. Nonetheless, no convincing case has yet been made to prove that senescence is an active, endogenously regulated process leading to the death of the whole plant in perennials."[37] We can note synonyms for exhaustion such as "premature decline" that are applied to forests, yet where premature exhaustion, premature decline, or early decline has been invoked, it is usually in the context of investigations of pests or disease causes,[38] and the same is probably true for grapevines. Historically, exhaustion may have been invoked simply when plants looked or performed poorly.

Woody perennials are a rather odd life-form, like a bunch of dead cells with a skin of live ones. Many woody perennials, including grapevines (fig. 20),[39] can live for hundreds of years (table 4), and we know

FIGURE 20. CLAIMS TO OLDEST LIVING GRAPEVINE. There are various claims to the oldest living grapevine, with ages that range from 250 to 400 years old. This vine in Maribor, Slovenia, is said to be more than 400 years old.

that both native and cultivated vines survive without any pruning. In fact, "minimal pruning" is a creative, successful, and labor-saving vineyard management technique that leaves hundreds of buds instead of the conventional tens of buds.[40] The success of minimal pruning is a serious problem for the notion that vines must be pruned for their own good.

"Premature exhaustion" was apparently more of a popular term that found little traction in crop studies, and never made its way into the scientific lexicon. There were a few scattered references to premature exhaustion of other crops, but the limited use of the term peaked in about 1850.[41] Yet, "exhaustion" was still invoked in winegrowing books, including Albert Winkler's classic 1974 text *General Viticulture*[42] and Hugh Johnson's 1989 *Vintage: The Story of Wine*, which claims that "vines must be pruned or die of 'exhaustion' from heavy crop."[43] Thousands of viable unpruned grapevines disprove this statement, and it is readily observed that grapevines get by just fine without human interactions; however, the exhaustion idea has had legs in the world of winegrowing.

We do not prune vines to keep them from growing themselves to death. Originally, pruning was probably practiced to keep vines in a shape that works well for farming. Without pruning, the new growth and fruit would move each season, in many cases upward (if a support like a tree was available for the climbing growth habit of the vine), which would be inconvenient come harvest time. In the cultivation of domesticated grapevines, we might add "reducing yield" to the list of objectives, if the vines produced more fruit unpruned than desired. Although this is the sense that one gets from popular publications in winegrowing, in most fruit crops, pruning is actually considered a yield-increasing practice.[44] Grapevines (and other crops) do not produce themselves to death, except perhaps under special conditions that involve other stresses. The bottom line for premature exhaustion is that the physiological factors determining the life span of grapevines and other woody perennials are almost a mystery, and it is well known that fungal wood rots, viruses, or market forces are what typically bring a commercial vine to its demise.[45]

THE BEGINNINGS OF VINE BALANCE IN FRUITFULNESS

Nevertheless, from this uncertain exhaustion idea—that the act of making fruit is injurious to vine health—it is a natural progression to consider what the consequence of too *little* fruit might be. (Something in between would be considered balanced.) Although one can sense in their writing that both Ravaz and Bioletti probably embraced a balance of fruitfulness, only Bioletti explicitly addressed this question of balance. The earliest description of a balance concept in viticulture was apparently introduced by Bioletti in 1897: "One of the chief aims of pruning is to maintain a just equilibrium between vegetative vigor and fertility."[46] Bioletti elaborated somewhat on his balance concept in his "Physiological Principles of Pruning," stating that "other conditions being equal, an excess of foliage is accompanied by a small amount of fruit, and an excess of fruit by diminished foliage," and advised that "a vine which has become enfeebled by over-bearing should be pruned for wood."[47] Bioletti had introduced vine balance by describing the equilibrium of fruiting with vegetative growth. If the vine grew too much vegetatively, it would not be fruitful. Bioletti's thoughts on equilibrium have a physiological basis in flowering and a potentially objective balance point that would maximize fruitfulness (flowers or clusters per vine or per shoot) at an intermediate point between low and high vegetative growth.

The fruitfulness-based vine balance is easily followed from Bioletti to his successor, Albert Winkler. In his 1897 "Vine Pruning," Bioletti wrote:

> Failure to reckon with this fact [a trade-off of vigor and fertility] and to maintain a proper mean between the two extremes leads, on the one hand, to comparative sterility, and on the other, to over-bearing and premature exhaustion.[48]

Nearly forty years later, Winkler wrote in "Pruning Vinifera Grapevines":

> Failure to reckon with this fact and to maintain a proper balance between the two extremes leads, on the one hand, to comparative unfruitfulness, and on the other, to overbearing, with poor quality of fruit and undue depression of the vine's capacity.[49]

Winkler repeats Bioletti's description almost verbatim. These quotes appear within "principles of pruning" sections first penned by Bioletti and revised by Winkler. Both authors were concerned with fruitfulness, referred to variously as fertility, sterility, (un)fruitfulness, and overproduction. Bioletti spoke to fruit quality frequently in his 1913 "Vine Pruning in California," and Winkler explicitly added that excess fruiting leads to "poor fruit quality" in his "principles of pruning." Like Ravaz's, these comments on quality refer mostly to sugar.[50]

A Balance of Vegetative and Reproductive Growth

Before focusing on fruit quality, let's consider equilibrium of fruitfulness and vegetative growth as described by Bioletti and Winkler. Winkler provided scant data supporting the concept—just the four data points shown in figure 21A, with the curve conveniently drawn showing an inflection (balance) point that could easily be dismissed for lack of evidence. However, Newton Partridge saw this fruitfulness response in Concord grapevines in Michigan and reported on it in the 1920s, calling the high yield point "optimum vigor."[51] In the text *General Viticulture*, Winkler cites a 1937 Australian publication as showing the same type of fruitfulness relationship to shoot growth—a curve with an inflection point.[52] Viticulturists, however, have not pursued the idea.

I have seen this phenomenon (high vegetative growth coupled with low fruitfulness) in heavily fertilized and irrigated grapevine and cotton fields. The fact that lush growing conditions lead to more vegetative growth and less flowers is well known in cotton, where it is referred to as "rank cotton."[53] A contemporary example can be seen in the variation in yield with shoot length of Thompson Seedless grapevines, where the differences in shoot length were created by various water applications (fig. 21B).[54] Australian research scientist Paul Boss made the observation that on rapidly growing shoots, tendrils (rather than clusters) usually develop.[55] Scattered over many decades are examples of conditions that when too poor, inhibit fruitfulness and when too lush, also inhibit fruitfulness.

What I am suggesting is that Bioletti was correct in that there is a physiological balance in fruitfulness that depends on the differential sensitivities of shoot growth and flowering to environmental conditions, and that this specific balance has a basis in empirical experience. This "balance point" is the maximum (inflection) in figures 21A and 21B. If correct, this principle of grapevine fruiting behavior lends itself to managing inputs, ostensibly to maximize fruitfulness, rather than managing crop by pruning or cluster thinning. Although vine balance as represented by fruitfulness has not been developed, the idea is sometimes implied in discussions of vineyard balance.

A Long-Term Stability?

Dry and colleagures cited Ravaz as the original source of vine balance with respect to both quality and consistent production. Of course, the latter would be true if a vine being out of balance meant death, but today one would have to really try to kill healthy grapevines.[56] This supposed balance could be further reevaluated as a long-term stability that may apply to the year in, year out fruiting and budbreak of grapevines. Both fruitfulness (the number of clusters and flowers on a cluster) and budbreak (the fraction of buds that break dormancy and grow in the spring) are highly variable year to year. I say "reevaluated" because it is my sense that this long-term equilibrium is also the same as what was meant by the original use of "vine balance," implied in some viticulture contexts, but not directly studied.

Pruning vines clearly disturbs their growth habit compared to what growth would have been without human interference. In trials at the Oakville Experimental Vineyard in California's Napa Valley, Cabernet Sauvignon vines were pruned to fewer and greater numbers of buds/vine than the standard Napa Valley practice. When yield was increased more than 50 percent in one season by leaving many more buds at pruning or by supplemental irrigation, the high yield was not sustained in

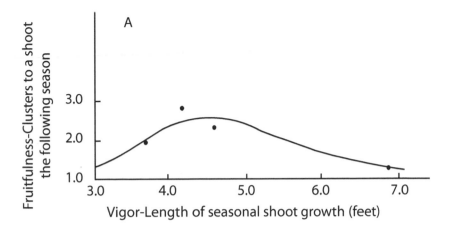

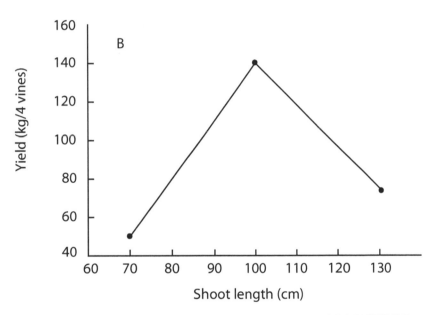

FIGURE 21. TWO EXAMPLES OF EXPERIMENTAL DATA THAT ARE CONSISTENT WITH THE ORIGINAL BIOLETTI CONCEPT OF A BALANCE OR EQUILIBRIUM BETWEEN REPRODUCTIVE AND VEGETATIVE GROWTH. (A) Winkler's data on balance, relating fruitfulness (defined as clusters per shoot) to "vigor" (defined as shoot length at the end of the season). (Redrawn from Winkler 1962, 229, fig. 79.) The data are from two varieties, Muscat of Alexandria and Alicante Bouschet. There are no further details to indicate what made the shoots longer, what was sampled to make the data points, or even which data points belong to which variety. (B) A similar relationship to that shown in A for Thompson Seedless, in which yield shows a maximum in relation to shoot length, and is reduced on vines with shoots shorter or longer than the length that corresponds to the maximum. Yield and shoot length were varied by water supply. (Data are from Williams et al. 2010a and b and are for the last of several years of the trial; each year the response increased toward the dramatic increase and decrease shown.)

the following season.[57] Similarly, the percentage of budbreak may show an inverse relationship to the number of buds left at pruning when the number of buds is high, as was observed in a pruning study by Markus Keller in Washington State.[58] Pruning to leave a high number of buds causes some buds to remain dormant that would otherwise have grown into a shoot, resulting in a lower percentage of budbreak. If vines are pruned to a low bud number, budbreak can reach more than 100 percent of the count buds (those left intentionally at pruning).[59] In between is a balance point. If vines are unpruned or largely unpruned in the minimal pruning system, the yield and shoot growth find these balance points and stabilize over time.[60] Regardless of whether this putative stability of budbreak improves fruit for winemaking, the hypothetical and testable "balance" is consistent with observations of vine growth and source-sink relations.

Fruit-Shoot Balance and Reproductive Cost

In 1906, Louis Ravaz wrote:

> Le raisin est un parasite dont la puissance dépasse de beaucoup la puissance de ceux avec lesquels les vignerons ont ordinairement à compter.

> The grape is a parasite whose power significantly surpasses the power of those that winemakers normally have to reckon with.[61]

All three early viticulturists (Ravaz, Bioletti, and Winkler) were concerned that allowing vines to produce fruit was detrimental to vine health, but Ravaz had a somewhat unusual take on the nature of the grapes that grow on grapevines. Parasitism is the *nonmutual* relationship between *two* organisms. Most biologists think that if there is a purpose to an organism's life, it is reproduction—the bearing of fruit for the grapevine.

Clearly, vegetative growth is required to produce reproductive growth (fruits and seeds). The Harvest Index (a unitless ratio of amount of harvested crop/total biomass, or amount of harvested crop/total above-ground biomass) is a metric that is widely used in agronomy. The Green Revolution of the 1960s was in part a consequence of selecting for grain varieties with a higher Harvest Index and reduced partitioning of photosynthate into stem or stalk height.[62] Current wheat genotypes have a Harvest Index of 0.45–0.6, and a couple of estimates for grapevines in the literature have been on the high end of that range. One can go only so far in reducing shoot growth (increasing the Harvest Index) because one must first have the source (leaves) in order to grow and ripen the grapes.

Beyond the Harvest Index, there is thought to be a cost of reproduction, an ecological concept that emerged in the 1980s. While this notion of internal competition is present not just in winegrowing but in biology as a whole, it has been somewhat controversial. Reviews of the topic have included comments such as the following: "However, despite the central importance of the cost of reproduction to theoretical models and explanations of patterns observed in nature, the empirical evidence to support this concept is weak";[63] "However, what appeared to be crystal clear in theory later became less evident when field biologists attempted to evaluate the cost of reproduction hypothesis."[64] Thus, although it is recognized that resources used for one kind of growth cannot be used for the other, it has been difficult to demonstrate the perceived cost of reproduction, particularly since reproduction is generally accepted as the primary plant objective.

An interesting study by Johannes Knops and colleagues, published in the *Proceedings of the National Academy of Sciences* in 2007, observed the reproductive and vegetative growth of California oaks over several seasons. Based on the responses to dry seasons, the authors showed that vegetative and reproductive growth in oaks have different programs for responding to the environment, and therefore do not simply compete for reserves: "We conclude that the observed negative correlations [between vegetative and reproductive growth] are not causal, but rather a consequence of growth and reproduction being dependent, in opposite ways, on environmental conditions."[65] Indeed, Knops and colleagues show that the negative correlations between vegetative and reproductive growth that are commonly used to support the trade-off concept are potentially misleading and clearly not sufficient evidence to conclude that a trade-off is in effect.

In the grapevine, it is similarly clear that vegetative and reproductive growth do not have the same relationships with the environment.[66] Vegetative growth is more sensitive to water deficits than reproductive growth, and the same is probably true for responses to nitrogen as well.[67] It is also commonly observed in some crops, and occasionally in grapevines, that the presence of fruit sustains leaves and the removal of fruit (including harvest) leads to accelerated leaf senescence. This is the opposite of the effect that one would predict if in fact fruit compete for reserves at the expense of vegetative growth. Furthermore, the balance concept that addresses fruitfulness based on allocation of resources for reproduction (nutrients and photosynthate) has no clear physiological connection to the other version, in which a balance of vegetative and reproductive growth is considered to be key to winegrape quality (except when too much fruit prevents adequate ripening).

THE RETURN OF YIELD: PRUNING WEIGHT IN FRUIT AND WINE QUALITY

The more prevalent version of vine balance in winegrowing today focuses on fruit quality rather than fruitfulness, although a conflation of the two has been common from early on. Of course, quality is the end result we are interested in; however, that is for a wine model of our choice today, which has historically been ephemeral. This should give us pause when considering the idea that there is a physiologically based balance between vegetative and reproductive growth that yields optimum fruit composition.

Early Efforts to Address Fruit Quality with Vine Balance

Ravaz's final reference to Y:PW is found in his 1911 work on leaf thinning, and references to Ravaz's work apparently ceased with the close of his career. Researchers and winegrowers of the time did not follow Ravaz's lead in using Y:PW as a cropping guideline, nor even as a research tool in the quest for a guideline.[68] The ratio apparently died or went dormant, perhaps with the new understanding of brunissure or with Ravaz's passing. Winkler worked mostly with table grapes, but in both table grapes and winegrapes he found that lower crop relative to growth of vegetative plant parts (whether evaluated as trunk growth, number of leaves, or pruning weight) correlated with higher sugar in fruit. Later in the 1950s, Winkler also showed that removing leaves decreased fruit size, Brix, and color in table grapes. These findings are similar to those of Ravaz. Winkler made contributions by developing quality scales for table grapes and winegrapes based on a combination of Brix and acid concentration; however, there is no balance point in vegetative growth shown or interpreted in his writings. Winkler seldom reported pruning weight and never Y:PW.[69]

The rediscovery of Y:PW actually commenced before the Ravaz Index emerged from that 1995 meeting in Italy. If credit is due for the return of Y:PW, it is in large part to the Israeli researcher Ben-Ami Bravdo and his coworkers, who published an influential series of five papers in the 1980s, some of which are cited by those who introduced the Ravaz Index in proceedings papers from the Conegliano meeting. Bravdo suggested the use of the Y:PW as an alternative to yield for evaluating how much crop to produce.[70] Several other leading viticulturists adopted Y:PW in the 1980s, including Stan Howell, Mark Kliewer, and Richard Smart, all without citing Ravaz, and hence evidently without knowledge of Ravaz's use of the ratio seventy-five years prior. Given that yield is always measured and that pruning weight has been commonly used as a measure of annual vine growth, it is not surprising that such an easily obtained ratio was generated in field research multiple times, but kudos to Scienza and colleagues for pointing out that Ravaz was the first do so.

Bravdo and colleagues (citing Winkler) sought to match crop load to vine size, and in doing so address the HYLQ paradigm discussed in chapter 1. In setting up their study, they refer to a problem created by several research papers in which fruit or wine quality was *not* reduced in the face of increased yields:

> There is need for a measure which will clearly define overcropping, undercropping, or normal cropping.

> The effect of crop levels on the growth of the vines and on the quality of the fruit and the wines was not always found to be consistent [seven references were cited here]. Even within the same cultivar, higher crops were not always associated with lower quality. It may well be that various results reported were inconsistent because crop level alone is an insufficient measure for cropping.[71]

Thus, when the experimental results were not consistent with HYLQ, it was apparently assumed that something was measured incorrectly, or that a better measurement was needed. The alternative explanation, that yield was not an important factor in those studies, was evidently not considered. Regardless, their work was important aside from any insight gained into the vine balance concept. One of the keys to the importance of Bravdo's work is the focus on wine quality, in the context of seeking the putative balance, rather than leaf browning, premature exhaustion, poor flowering, Brix, and so on. The importance of this switch toward making and tasting the resultant wines is difficult to overestimate. Most of the viticulture work up until this point compared fruit yield, sugar concentration, and sometimes color in reds. In trying to elucidate vineyard practices for better wines, the incorporation of objective sensory methods is all but essential, because we do not know what could be measured in the fruit that predicts how the wine will taste.

Another key aspect of Bravdo's work is that all fruit were at a similar maturity (21.5–22.0 Brix). As such, any differences in the rate of ripening caused by cluster thinning were largely eliminated as a potential source or cause of any wine differences. This is a very important distinction from the common approach of harvesting fruit on the same date. By harvesting at a sugar concentration, they say in effect, "Okay, we know that higher crop delays ripening," and move on to investigate whether or not that is the whole story.

THE RECOMMENDED VALUES FOR VINE BALANCE

The scientific, agricultural extension, and popular publications and websites that recommend cultural practices to growers consistently adhere to a vine balance concept that is expressed in Y:PW values of 5 to 10 in order to grow good grapes.[72] Bravdo's group reported that Y:PW ratios of 4 to 10 in Cabernet Sauvignon "covered a range where a positive correlation with wine quality seems to exist."[73] They supported the selection of the upper limit of 10 with examples from the literature in which quality was lower beyond that value, but in almost all of those studies there was no effect on wines for Y:PW ratios below 10. This series of studies is nevertheless the primary origin of the standard metric for vine balance.[74]

However, if we pay close attention to the relationships between wine scores and the Y:PW values that are presented in Bravdo's key studies, the situation is not as clear as the widespread agreement on 5 to 10 suggests. When the mean wine scores are plotted to show their dependence on Y:PW (which was not done in the original publications), the first thing to note is the small range of wine scores over a large range of Y:PW (fig. 22A). Whatever else these data may indicate, they show a low sensitivity of wine quality scores to Y:PW (similar to most of the research results with varying yield in the previous chapter). Next, note that the data segregate according to variety along the Y:PW axis: Cabernet Sauvignon has a Y:PW range of 3.0 to 6.4, and Carignane has a Y:PW of 9.6 to 19.3. The Cabernet Sauvignon wines showed no change in quality scores over that range of Y:PW, despite increasing the yield from about 8 to 18 kg (18 to 40 lb) per vine; the Carignane shows a significant drop in quality but only when the load was increased to nearly 20 kg (44 lb) per vine.

To their credit, the Bravdo team experiments were conducted repeatedly over several seasons. To take a closer look, all wine scores and Y:PW

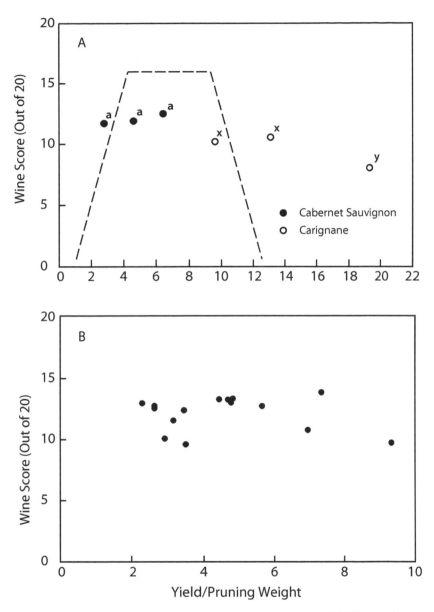

FIGURE 22. THE RELATIONSHIP OF WINE TASTING SCORES TO Y:PW. (A) Wines made from Cabernet Sauvignon and Carignane vines grown in Israel with various Y:PW, which was varied by flower cluster-thinning treatments at bloom. Unthinned vines resulted in 60–70 clusters/vine, and thinned vines with nominally 40/vine and 20/vine. (Data are five-year means from Bravdo et al. 1984, 1985a and b.) (B) Wine tasting scores at various Y:PW showing all data for wine scores for Cabernet Sauvignon at various Y:PW. (Data from Bravdo et al. studies cited in the text.)

TABLE 5 LINEAR REGRESSION COEFFICIENTS (R2) INDICATING THE
RELATIONSHIP OF YIELD TO Y:PW RATIO FROM SEVERAL EXPERIMENTS

r^2	Data from
0.994	Chapman et al. 2004a, pruning, two years of data
0.979	Ravaz 1904a, no treatments
0.953	Bindon et al. 2008, pruning to 30, 60, 120 buds/vine
0.955	Nuzzo and Matthews 2006, cluster thinning, two years of data
0.808	Keller et al. 2008, cluster thinning (p < 0.015), five years of data
0.569	Keller et al. 2008, irrigation treatments (p< 0.05), five years of data

from each Cabernet Sauvignon cluster-thinning treatment in each of the
five years of those studies are plotted in figure 22B. This expands the
range of Y:PW in Cabernet Sauvignon to about 2 to 10, but still reveals
no decrease in wine score at the high or low end of that range. From this
analysis, changes in yield and Y:PW, each more than 100 percent increases
from low to high, caused essentially no detectable change in Cabernet
Sauvignon wine quality.

Y:PW OFFERS LITTLE NEW INFORMATION

Recall that Scienza and colleagues, who helped jump-start the Ravaz
Index at the meeting in Conegliano, were on an appropriate track when
they evaluated several means of measuring the putative vine balance.
What has been lost since then is that Y:PW came up lacking compared
to other measures in that study. There are several straightforward reasons why Y:PW fails to provide adequate insight into winegrape quality.

First, evaluating fruit at various Y:PW is very similar to evaluating
fruit at various yields. The most common means of altering "balance"
(i.e., Y:PW) are pruning and cluster thinning—the same methods as for
manipulating yield. In our Napa Valley yield trials (described in the
previous chapter), Cabernet Sauvignon vines were pruned to fewer and
greater numbers of buds/vine than the standard Napa Valley practice.[75]
For those vines, yield and Y:PW tracked each other over the approximately fourfold range of buds per vine (fig. 23A). In pruning experiments with Shiraz in Australia, yield and Y:PW varied similarly over a
wider range of yields and buds per vine (fig. 23B). In fact, for many
studies, there is such a close correlation between yield and Y:PW that it
is difficult to envision how the latter could provide new insight (table 5).
The very close relationship—when yield goes down, Y:PW goes down

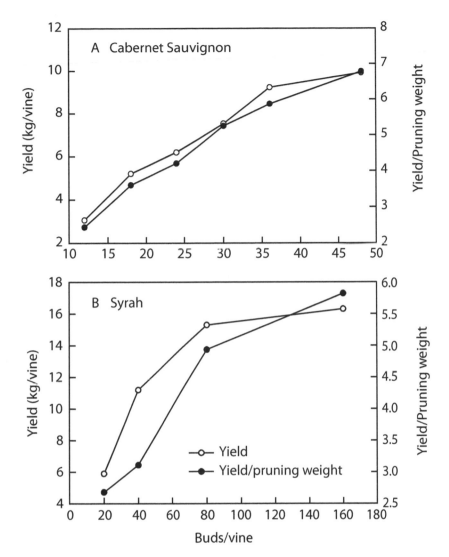

FIGURE 23. THE RELATIONSHIPS OF YIELD AND Y:PW TO THE NUMBER OF BUDS/VINE LEFT AT PRUNING. (A) for Cabernet Sauvignon grown in Napa Valley, California (data from Chapman et al. 2004a and b), and (B) for Shiraz grown in Griffith, Australia (data from Freeman et al. 1980).

similarly—means that one could be substituted for the other in most cases.

Y:PW changes in a similar manner to yield, because when one prunes a grapevine to have more or less buds, the vine produces more or less crop; with each reduction in bud number, there is an approximately corresponding reduction in cluster number. The major vegetative adjustment is that with fewer buds (shoots), each shoot grows longer and therefore weighs more, leaving the total weight of shoots relatively unchanged. This has been evident in various reports for some time, such as in the data from a 1970s pruning study in California, where pruning to a high bud number in consecutive seasons increased yield by about 65 and 50 percent, but reduced pruning weight by only about 16 and 10 percent.[76] Although pruning weight may increase slightly when clusters are removed by thinning, the potential to affect vegetative growth is low, and decreases as the season progresses. In another example from Australia, when vines were pruned to more buds, yield increased substantially; however, "pruning level had no effect on pruning weight of either irrigated or nonirrigated Shiraz."[77] It may seem counterintuitive, but pruning weight is relatively insensitive to changes in pruning within some range of buds retained. At very high bud numbers, yield stops increasing, but so does the pruning weight (and in some cases it may actually decrease).[78]

Because yield and Y:PW tend to track each other when changed by thinning and pruning, aspects of fruit ripening can be expected to be similarly related to either. The close correlation of yield and Y:PW has been in the published data for a long time, but has not been acknowledged. In fact, Ravaz's data also show changes in yield that are very similar to changes in Y:PW (table 5). Accordingly, we get the same picture in the corresponding relationships of fruit ripening to yield and to Y:PW where there is a clear inverse relationship between must density and Y:PW (fig. 24A), and between must density and yield (fig. 24B). Ravaz was explicit in pointing to the Y:PW ratio as being more important than the yield, yet was clearly aware that there was a relationship with yield that was similar to that with Y:PW. In presenting the data for figure 24 he said: "So the quality is declining as production increases following the law that I have established [in my work on brunissure]. For large productions obtained with a large number N of clusters, it decreases further as N increases."[79]

In his 1904 brunissure book, Ravaz showed that the wine alcohol concentration (which is directly dependent on fruit sugar concentra-

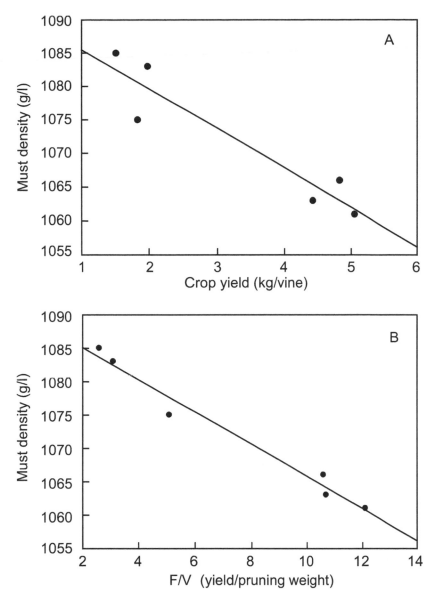

FIGURE 24. RAVAZ'S DATA ON FRUIT QUALITY AND Y:PW (F/V IN RAVAZ'S NOTATION) SHOWING FRUIT QUALITY AS HAVING A SIMILAR RELATION TO YIELD AND TO Y:PW. (A) Must density (sugar concentration) at various crop yields. (B) Must density at various Y:PW for the same vines as in A. (Data from Tables 34 and 35 in Ravaz's 1909 *Cultural Operations* booklet.) In his 1904 brunissure book, Ravaz showed that the wine alcohol concentration (which is directly dependent on fruit sugar concentration) is also closely and inversely related to Y:PW (shown in fig. 19).

tion) is also closely and inversely related to Y:PW (fig. 19). Ravaz's data on wine alcohol concentrations might justify a reference to his use of Y:PW as a predictor of fruit quality (or how much crop to keep) when sugar accumulation is the main concern for fruit quality.[80] The close relationships of wine alcohol to Y:PW and must density (i.e., sugar concentration) to Y:PW are not revealed in other Ravaz data sets for soil color, timing of pruning, and direction of shoot orientation. Just as we saw with yield and berry size in the previous chapter, these inconsistencies are signs that Y:PW is related (perhaps indirectly) to sugar accumulation sometimes, but not in a causal way.

A second reason for the failure of vine balance (as indicated by Y:PW) to offer new insight into how to best grow winegrapes is that Y:PW seldom falls outside the recommended 5 to 10 value for most winegrape varieties under most conditions. The original Bravdo data for Cabernet did not; the Napa Valley study of cluster thinning in Cabernet Sauvignon, which left two times more buds than usual (4-bud spurs instead of 2-bud spurs), resulted in Y:PW values of 5.85 and 4.37 in consecutive seasons,[81] and leaving 20 to 160 buds per vine in Shiraz resulted in Y:PW ratios of 2.68 to 5.82.[82] The main concern from Ravaz to Bravdo to today has been too much yield. However, with the target range including a 100 percent increase from 5 to 10, the fact that essentially all of that increase is typically in yield, and that Y:PW above 10 is rare, makes the utility of the target uncertain.

A third reason for the failure of vine balance is that the use of Y:PW to measure balance incorporates both the assumption that pruning weight is a good indication of the leaf area and the supply of photosynthate for berries. But it is not. In plant source-sink parlance, the pruning weight is assumed to represent the "source" to the fruit "sink" (although the wood serves as both sink and source). Bravdo and colleagues cite several studies in order to argue that pruning weight is a good estimator of leaf area, but evidently did no measurements themselves to demonstrate the concept. In a strange parallel to the studies citing Ravaz for having shown a strong correlation between leaf area and pruning weight, I found that the cited studies in the Bravdo work do not bear out their claim.[83] In fact, one of the cited studies on the timing of leaf removal reported the opposite: "These data indicate that pruning weights of canes would not be a good indicator of reduced vine capacity due to loss in leaf area, especially if defoliation occurred at veraison or later."[84]

When a significant positive correlation has been reported between leaf area and pruning weight (and a positive relation *is* expected by the conventional wisdom), it has seldom been a strong (i.e., an accurately predictive) relationship.[85] Two studies, one conducted in California and another in South Africa, show a tight correlation.[86] However, those may be fortuitous because there are many variables that affect the relationship of leaf area to pruning weight, such as internode length, the fraction of leaves borne on primary versus lateral shoots, and confounding factors such as shoot thinning, leaf thinning, and so on that affect leaf area more than pruning weight. A rough relationship like those shown in figure 25 is representative. There is a positive relation, but it is not close enough for pruning weight to be a good predictor of leaf area, especially when the relationship is very different for Cabernet Sauvignon grown in Paso Robles when compared to that grown in Oakville, California (fig. 25).

Pruning weight is a rough but poor metric for leaf area, but then leaf area is not a critical thing to know—unless it is known that leaf area determines winegrape quality. Although seldom tested, pruning weight does not fare well in predicting fruit composition, which should be the whole point of the exercise. As Italian viticulturist Stefano Poni observed, "The data (on sugar concentration, anthocyanins, phenolics, pH, and K+) also pointed out that the differences among trellises could not have been predicted simply on the basis of widely accepted indicators of crop load (e.g. the yield-to-pruning weight ratio) or canopy density (e.g. leaf area-to-canopy surface area)."[87] Indeed, the same Y:PW value cannot have the same meaning for the vine's source-sink relations if arrived at by cluster thinning at veraison versus winter pruning versus various degrees of leaf thinning or shoot positioning.

The direct source–sink relations of leaf area to fruit ratios was investigated in a series of studies by Mark Kliewer at UC Davis.[88] Kliewer's work showed that berry growth, Brix, and red color all reached maximum values at about $10-12$ cm^2 leaf area $\times$ gm^{-1} fruit. While it is clear that some amount of leaf area is needed to grow and ripen a berry ($10-12$ cm^2 leaf area/g fruit is currently the best estimate), the relationship of leaf area to fruit ripening is also weak. That should not be surprising. Photosynthesis is an environmentally sensitive process and is not simply determined by leaf area. Conditions may occur in which photosynthesis is inadequate for ripening for reasons other than leaf area, including shading, disease, water or nutrient deficits, and so on. Results from a study on Grenache in southern France "show that berry composition is less sensitive to leaf:fruit ratio than to grapevine water status, especially when the lower limit in terms of leaf:fruit ratio has not

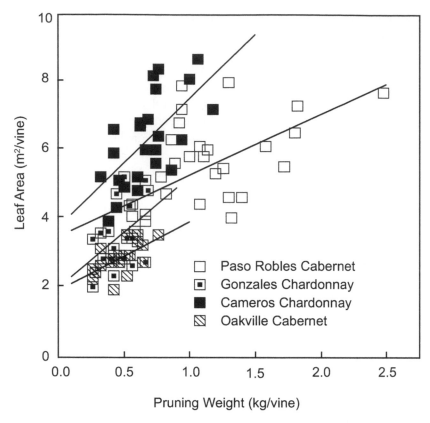

FIGURE 25. THE RELATIONSHIP BETWEEN VINE LEAF AREA AND VINE PRUNING WEIGHT
FOR TWO CHARDONNAY AND TWO CABERNET SAUVIGNON VINEYARDS IN CALIFORNIA.
There is no single relation that would allow vine leaf area to be predicted from Y:PW.
For example, for a vine with 1 kg pruning weight, the vine leaf area ranged about 100
percent, from 3.5 to 7.0 m²/vine. (Data courtesy of L. E. Williams.)

been reached."[89] In other words, leaf area (and therefore Y:PW) can be
helpful when approaching insufficient leaf area to ripen the crop. Furthermore, recently produced photosynthate is allocated to various sinks,
including roots and wood in the above-ground vine, not just the fruit;
thus, the source-sink relations of fruit ripening are more complicated
that simply relating leaves to berries.

There has been some recognition of the limitations of Y:PW. Recently,
viticulturists have tried using pruning weight (as well as yield and other
vine size parameters), expressed per unit length of row, in order to realize their anticipated relationships. For example, when Kaan Kurtural
and coworkers at Fresno State University in California imposed pruning

treatments and reported yield/meter of row, the relationship of yield/meter to Y:PW was very close (r^2 = 0.97),[90] similar to those in table 5, where yield was reported per unit area. These metrics are all the same values as before, multiplied by a constant that depends on row spacing, and thus have little potential to create insight. More informative may be shoot density, which although indirect, is nonetheless somewhat predictive of the cluster microclimate.[91]

CONCLUSIONS

Viticulturists have been largely circumspect on HYLQ, and perhaps thought that they had found a more appropriate metric for fine wine-growing in the vine balance concept. Despite the obvious need for some amount of leaves to develop grapes, the well-documented clarity in the trade-off of fruit size and number and in root-shoot balance has not been realized in vine balance. In retrospect, this should not be surprising.

Any balance metric designed to bring potential "yield effects" on quality into line was destined for difficulty from the outset, because yield and balance (expressed as Y:PW) are usually closely correlated, and the HYLQ paradigm is poorly supported by evidence. The close correlation of yield and Y:PW tells us that Y:PW is unlikely to be a significant improvement over yield as an index for fruit or wine quality. While HYLQ predicts an ongoing decrease in quality with increasing yield, vine balance predicts a balance point or optimum fruit quality at some Y:PW. The responses of fruit or wine parameters to Y:PW have not been shown to change in a way that reflects an optimum (increasing and then decreasing), which should be revealed if in fact a balance point existed. Neither the early works of Ravaz nor modern Y:PW data reveal a consistent balance point in fruit sugar, color, or other quality parameters that would make for an effective production objective. In general, there is no "other side" of the balance where too much pruning weight or leaf area inhibits or otherwise negatively impacts fruit ripening, although this may occur with the old balance idea regarding fruitfulness (fig. 21).

What we do see in studies since Ravaz is that lower Y:PW (or yield:other estimates of vegetative growth such as leaf area) results in higher or earlier concentrations of sugar (or color, etc.). We find ourselves at a familiar place. This situation is the HYLQ concept as it is most clearly and consistently seen in the grapevine—in the role that crop load plays in the timing of ripening; lower yield or Y:PW leads to riper fruit in limiting environments. Add to these issues the fact that for many vineyards, it is difficult to reach levels outside the recommended balance of Y:PW 5–10, and it's clear that the attractiveness of a number that was convenient to measure may have unfortunately led to its hasty adoption. In some cases, such as the high-yielding Carignane in Bravdo's studies in the 1980s, quality parameters or wine scores drop off at high Y:PW ratio—just as they do when yield is employed as the basis of quality (as depicted in the hypothetical curve 2 in fig. 6). Dry and coauthors suggested that using earliness (Brix) or other maturity measures is more important and useful than a Y:PW metric, and the analysis in this chapter supports that suggestion.[92]

The grapevine may be unique in the fruit that it produces, but it necessarily shares many attributes with other crop plants. Vine balance is omnipresent in winegrowing, appearing in a wide array of publications, from academic research to extension guides to winery websites and, to a lesser extent, in the wine press,[93] yet no concept parallel to vine balance is found in other crops. This should give us pause. If the viticulturists had discovered something important in grapevine physiology, it would have been picked up by researchers or producers within other crop specialties. An important problem for the vine balance concept (among others in winegrowing) is that it is foremost an issue of faith, a priori reasoning that there is a balance of leaves and clusters that will result in wine attributes matching the current human model of what a wine should taste like. The mind-set that there must be a balance point of leaves and fruit that produces the finest grapes and wines continues despite, rather than because of, the empirical record. (As does the endless repeating of experiments, in which leaves and/or clusters are removed with little or no refinement.) It is telling that as recently as 2014, publishable experiments were still being conducted in which one-third of the leaves or one-third of the fruit can be removed in an effort to learn how to manage the crop,[94] and vine balance is described as an "appropriate relationship between vegetative and reproductive growth" but not defined further.[95]

It might seem like viticulture heresy to question what is meant by the term "vine balance"—after all, who could be against balance? Despite much verbiage, however, the basis of the concept remains shaky, as does its utility in winegrowing. As L. G. Santesteban and coauthors put it recently in their work "Vegetative Growth, Reproductive Development, and Vineyard Balance," the balanced vineyard is the most frequently cited goal in winegrowing; "however, when further asked, they (we) fail to define precisely what a balanced vineyard is."[96] Indeed. Vine balance is a belief, looking for a justification in the physical world.

Just as with the HYLQ and BBB myths, it is important to realize that obtaining good fruit is not as simple as reducing yield or balancing leaves and clusters. The source-sink relationships that metrics like Y:PW attempt to describe are but a part of the overall environment that impacts final winegrape quality. The oversimplified source-sink perspective dominates to the detriment of investigations into other important factors, such as vine water status and fruit microclimate. In the late twentieth century, viticulture researchers Richard Smart and Mark Kliewer were important in directing attention to the cluster microclimate—the light, wind, humidity, and temperatures that affect fruit growth and ripening. But the Y:PW does not describe microclimate, and therefore cannot predict the sensory characteristics such as vegginess of MIBP, various phenolics that affect color and astringency, and so on that are strongly affected by the berry environment. This is being revealed in the vineyard where "substantial increases in crop load (Y:PW) seem to have little impact on wine sensory quality unless accompanied by major increases in canopy shading."[97] And in a study in Perugia, Italy, that employed a creative new design that allows the trellis to move, and hence change the cluster microclimate, Y:PW was ineffective in predicting fruit composition.[98]

Viticulturists refer to vine balance when they are trying to restrict what they consider excessive vegetative growth—growth beyond what is needed to ripen the crop, or what they consider excessive crop, that is, too much fruit for the grapevine to ripen properly. Plant physiologist and eminent wine expert Maynard Amerine said that "wine quality is much easier to recognize than define," and the same seems true for vine balance.[99] Vineyard managers can develop an eye for a good vineyard, in which their experience helps integrate more than leaves and clusters when visually estimating "balance," by which they may really mean their evaluation of the whole production system—including crop load, cluster microclimate, as well as a background sense of the soil and climate restrictions on productivity and ripening. Interestingly (although this relates more to the aesthetic aspect of balance briefly mentioned at the beginning of the chapter), this approach to evaluating vineyard balance may be more similar to how the grapevine actually goes about its business—integrating the many signals from its environment. In the next chapter, we will review the conventional wisdom surrounding some of the additional environmental factors beyond leaves and fruit.

CHAPTER 3

Critical Ripening Period and the Stressed Vine

The conventional view serves to protect us from the painful job of thinking.

—John Kenneth Galbraith

"Critical ripening period" and "physiological maturity" are phrases used by winemakers that appear frequently in conjunction with winegrape harvests, on winery websites, and in wine press reviews of vintages, winegrowing regions, and wines. Both phrases also have histories in crop science and production. "Critical ripening period" has long had a vague meaning in discussions of other crops, including raspberries, citrus, dates, coffee, and even commodities such as rice and cotton.[1] "Physiological maturity" is also used in agronomy and horticulture, but with explicit definitions that refer to a time before readiness. In winegrowing, winemakers want to harvest at physiological maturity, and have claimed that deciding when to harvest the grapes is the "single most important viticulture decision."[2]

The ripening period of grapes (see fig. 2) is appropriately given special attention in the production and research of grapevines, because that is the time during which much of the action takes place in the fruit. This emphasis brings together the "critical" aspect of the critical ripening period and the objective of the fruit reaching full physiological maturity. In addition, the *rate* of ripening is widely considered to be an important variable of winegrape development and ultimate quality. Most popular writing holds that slower development is better, asserting that ripening can in fact proceed "too fast."[3] This chapter will discuss the rationales and evidence bearing on these concepts. We will also look briefly at how temperature, light, and water may affect the rate of ripening. This discussion leads naturally into a consideration of another popular myth—the idea that the "stressed vine makes the best wine."

THE WHOLE SEASON AND DEGREE DAYS

It is critical to have sufficient heat as well as frost-free days in a season in order to develop and ripen a crop. Winegrape varieties vary somewhat in their minimum heat requirements, in rates of development from budbreak to harvest, and in their responses to environmental cues. In cool climates, heat and the length of the growing season (from frost to frost) are the constraints that determine what varieties can be successfully grown and ripened.

Available heat impacts the quality of the vintage, in part because obtaining a sufficient ripeness (Brix) may not be ensured if grapevine development is slowed by cool weather. In general, cool temperatures slow all aspects of winegrape development from budbreak to harvest; the conventional (or "rule of thumb") time to harvest historically has been 100 DAA (Days After Anthesis) in Champagne and 110 DAA in Bordeaux.[4] Those standards incorporate the use of adapted varieties that complete crop development within the normal growing season of their respective climates. Winegrowers in Burgundy found genotypes (Pinot noir, for example, an early and slightly shorter season variety), cultural practices (such as low yield), and a wine style (creating wine from grapes that are less ripe than is possible in warmer climates, which today is sometimes called a "finesse" wine) that fit the limited ripeness that has been possible within the constraints of the climate. In Champagne, where ripening is perhaps even more challenged, early varieties and a different wine style fit the environmental constraints. For Champagne (and other sparkling wines in the same style), grapes are harvested at lower sugar than that needed for good still (not bubbly) table wine, and sugar is added in a second fermentation. A cooler climate still? *Vitis vinifera* grapes, which are not cold tolerant, may not be adapted. Thus, heat is a major limitation in the distribution of wine-growing regions and varieties.

Heat summation, which is any of various calculations that sum air temperature over time, is one means of characterizing climate and weather. Although not common in the popular press, viticulturists and academics use heat summations to determine the appropriate variety for a site and the anticipated rate of ripening. One example can be found in the work of Albert Winkler: "The time of ripening is determined primarily by variety and heat summation."[5]

The most common metric for heat summation is the degree day, or growing degree day (GDD), in which a minimum threshold temperature is selected below which no vine development is either observed or expected. For grapes, this threshold temperature is 50°F (10°C) by convention.[6] In classic studies conducted in the 1930s and 1940s, Maynard Amerine and Winkler used accumulated GDD measures during the season (April through October) to designate winegrowing regions and to formulate recommendations for the growing of winegrape cultivars in California.[7] They set five regions, Region I being the coolest in terms of seasonal GDD, and Region V being the warmest. Their work has been highly influential and widely cited in viticulture research, used in some contract negotiations in which being able to show grapes are produced in a cooler region can be an advantage, and making its way into the work of some wine journalists. James Halliday includes the GDD as one of the characteristics of wine regions in his *Wine Atlas of Australia.*

The use of heat summation in grapevine phenology is sometimes attributed to Winkler,[8] but Winkler's work was hardly the beginning of heat summation in crops. The idea that plant development is dependent upon (and could be predicted by) heat summation was apparently introduced in France in 1735 by René Réaumur, who invented a thermometer and established a temperature scale based on his correlations of temperature summations and changes in plants, such as budbreak and flowering.[9] Réaumur summed the daily air temperatures in spring and "found the sum to be a nearly constant value for the development of any plant from year to year."[10] Réaumur assumed that his thermometric constant expressed the amount of heat required for a plant to reach a given stage of maturity, and this gave rise to the growing degree day unit of today. He was probably correct, or at least roughly so, as correlations between heat summations and plant development have been made with varying degrees of success ever since.

Winkler's work in the 1940s was preceded by that of Bioletti, who in 1915 used measures of average temperature, temperature summation, and precipitation to describe six viticulture regions in California.[11] Bioletti ascribed differences among regions based only on table grape, raisin, and wine uses. He lumped all coastal valleys together as the best region for winegrowing, and reported no measurements of fruit or wines. Winkler and Amerine's later work was much more comprehensive, intensive, and quantitative than Bioletti's cursory partitioning of climates; their undertaking was almost monumental in scale.[12] They analyzed fruit from many varieties grown in many locations around the state. Winkler and Amerine also made, analyzed, and tasted wines to guide their recommendations. Their system was widely acclaimed for pointing to the important role of climate in winegrowing, for quantifying the climatic effects, and for guiding the planting of varieties for decades. This system of classifying vineyards became the New World paradigm for winegrowing and was widely used in viticulture research around the world.

Another climate scheme for classifying vineyards and varieties that has received use among European academics is the Huglin Index, similar to Winkler's degree days with the addition of a parameter for the effect of latitude on day length. There is also some difference between the two systems in the definition of the beginning and end dates of the growing season, as well as in the definition of maturity as applied by Huglin and Winkler. Whereas there are no explicit minimum GDD values for specific varieties in Winkler's work, Pierre Huglin based his index on the minimum sum required to bring a given varietal to a sugar content of 180–200 grams per liter, which roughly corresponds to only 17–18.5 Brix (table 6).[13] Nevertheless, Winkler's GDD and the Huglin Index are closely related ($r^2 = 0.98$), which indicates that there is essentially no difference between the two numerically.[14] It is important to note that practical experience supports the general differences in heat requirements described by Huglin.

If heat accumulation determines the rate of development for each variety, there should be some fixed sum of GDD (or heat units) that is required to get the job done. By and large, this has been supported by practical experience. Pinot noir, for example, can be grown where the growing season is shorter and has lower GDD than that required to ripen Cabernet Sauvignon. Nominal GDD limits for many varieties are in place and are widely used for identifying adapted varieties in the United States, Canada, and elsewhere, especially in cool or continental climates where efforts are under way to develop new, adapted varieties.

Geographer Graeme McIntyre, along with viticulturist colleagues at UC Davis, conducted a multiyear, comprehensive study of the dates and heat summations for budbreak, flowering, and fruit maturity for many winegrape varieties grown in Davis, California, as well as for five important varieties grown in each of the five Winkler Regions of California.[15] The expected and observed responses of Pinot noir fruit ripening in the various regions are shown in figure 26 as an example. Instead of

TABLE 6 MINIMUM HEAT UNITS FOR SELECTED WINEGRAPE VARIETIES UNDER
THE HUGLIN INDEX

Huglin Index*	Winegrape varieties
1,500	Müller-Thurgau, Portugais bleu
1,600	Pinot gris, Gamay, Gewürztraminer
1,700	Pinot noir, Chardonnay, Riesling
1,800	Cabernet franc, Blaufränkisch
1,900	Cabernet Sauvignon, Chenin blanc, Merlot, Sémillon
2,000	Ugni blanc
2,100	Grenache, Syrah

* Lower limit around 1,400

reaching maturity in constant GDD and in shorter and shorter time (days) in increasingly warmer regions (as expected), the time to maturity decreased only from Region I to Region II (as shown in fig. 26B). From there on, the time required to reach maturity remained the same, but the GDD required increased. All varieties behaved in the same manner. This shows that the rate of crop development in California vineyards was limited by temperature (heat in GDD) in only the coolest areas (Region I, and perhaps Region II, in the Winkler and Amerine classification system). In regions warmer than Region II, the fruit ripened in the same amount of time (days), despite developing in increasingly warm conditions.

The Amerine and Winkler GDD system has been criticized for its oversimplification and inaccuracies. There are ample cases where the degree-day system has been demonstrably inaccurate in predicting timing of phenology (the timing of budbreak, flowering, veraison), and especially of fruit maturity as harvest approaches. Fair enough, but it is the uncritical applications of the work of Amerine and Winkler (or the GDD concept in general) that are the more appropriate target. It has long been recognized that plant growth and development are not so simply dependent on heat accumulation, and that other environmental factors, such as light intensity, must be taken into account.[16] In 1960, Jen Yu Wang, an expert in heat and crop development, reviewed the use of heat units in crop development, and found that "the heat unit system was widely adapted because of its value in satisfying practical needs, rather than for its accuracy or its theoretical soundness."[17] He pointed out several significant problems with expecting heat accumulation to predict crop maturity that apply to vineyards, including the following:

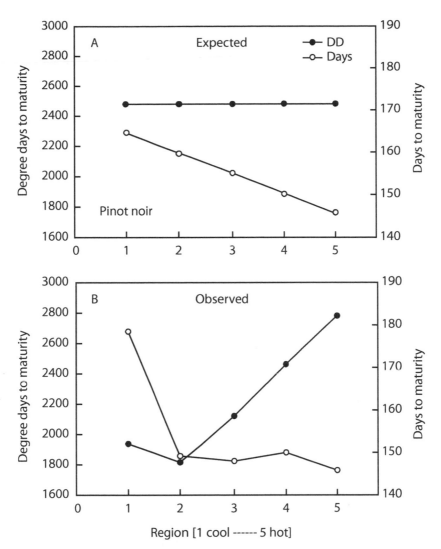

FIGURE 26. THE TIME REQUIRED TO DEVELOP PINOT NOIR WINEGRAPES TO MATURITY IN EACH OF THE FIVE CLIMATE REGIONS IN CALIFORNIA, ESTABLISHED BY ALBERT WINKLER. The time to maturity is given in two metrics of time: degree days (left axis) and days (right axis). (A) The expected time to maturity based on the assumptions that (1) the amount of heat (total degree days) required is constant for a variety, and (2) the rate of development is proportional to available heat. (B) The observed degree days and days to maturity. Pinot noir budbreak was about ten days earlier than Cabernet Sauvignon, but these relations were the same in each of the five varieties studied. (Data from McIntyre et al. 1987.)

1. Plants respond differently to the same environmental factor during various stages.

2. Growth response [to heat] may be linear only in the middle and is sigmoidal overall.

3. Heat units don't take into account temperature extremes [that affect metabolism in ways other than rate of development].

4. Heat units are not based on tissue temperatures [thus ignoring microclimate effects on berry temperature].

Wang's straightforward observations made more than fifty years ago tell us that there are about as many reasons to expect heat summation to *not* accurately predict grape maturity as there are to support it as a predictor. Yet alternate climate schemes for vineyards and varieties continue to come and go, without addressing these issues.[18]

As McIntyre and coauthors say in their introduction, "If temperature conditions are adequate, other factors such as solar radiation, day length, or soil moisture, may be more important limiting factors and are likely reasons for observed variations in accumulated degree day total."[19] Accordingly, the eminent plant physiologist Bryan Coombe repeatedly published studies in which his data and interpretations made this same point.[20] Many viticulturists continue to evaluate grapevine development and fruit ripening based on the same flawed assumption that the rate of development is driven by GDD in a simple way that just increases with more heat, but it is at least a little more complicated than that. Although at low temperatures heat will limit the rate of development, a key question remains—at what temperature do winegrape varieties reach a maximum rate of development?

TEMPERATURE AND THE RATE OF RIPENING

Winemakers and the wine press seldom rely on GDD to tell them about the rate of vine development, but together with academics, they do generally abide by the notion that fruit ripen faster at warmer temperatures. This is true in general, but the question arises whether the rate of ripening simply keeps increasing with temperature, or if it reaches some maximum rate beyond which it no longer increases or is even possibly inhibited.

Several lines of evidence suggest that grapevine development may be going as fast as it can at moderate temperatures, such as those experienced in Winkler's Region II (recall fig. 26). In Geisenheim, Germany

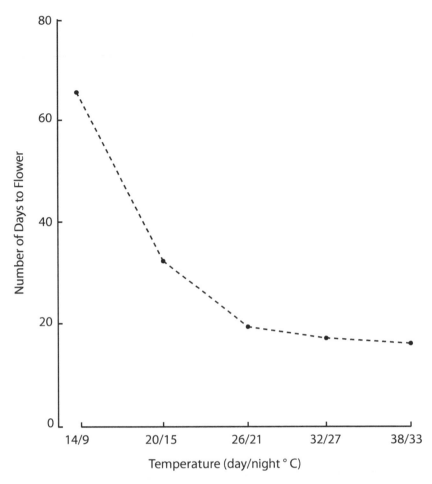

FIGURE 27. DAYS TO FLOWERING IN CABERNET SAUVIGNON AT VARIOUS GROWTH TEMPERATURES. (Redrawn using data from Buttrose and Hale 1973.)

(a Region I climate), Hans Schultz showed that the rate at which vines put on new leaves increased with temperature, but only up to moderate air temperatures of about 25°C.[21] In Australia, M. S. Buttrose and C. R. Hale conducted a series of groundbreaking experiments in the 1960s and 1970s using controlled environments. In one such experiment, several varieties were grown in different temperature regimes, and the time required to progress to bloom was recorded. The time from budbreak to bloom depended on temperature in a somewhat complex way, as shown in figure 27. Compared to vines developed in the coolest regime, small increases in temperature caused bloom to begin much earlier. As

TABLE 7   DAYS TO FRUIT MATURITY FOR CABERNET SAUVIGNON
GRAPEVINES GROWN IN VARIOUS TEMPERATURE REGIMES

Growth temperature (day/night °C)	Days to maturity
18/13	107
25/20	89
35/30	112

SOURCE: Data from Hale and Buttrose 1974.

the growth temperature continued to increase, the effect of temperature on the time to bloom diminished. After the daytime temperature had risen to above approximately 27°C, there was no longer any shortening of the time to bloom; the temperature effect was saturated, and at higher temperatures the rate of development from budbreak to bloom could accelerate no more.

In another experiment, when Cabernet Sauvignon vines were grown under different temperature regimes, berry development from bloom to maturity was fastest at a daytime temperature of 25°C and slower at both 18°C and 35°C (shown in table 7).[22] In addition, the highest Brix observed in the study occurred in fruit that developed at high temperatures during Stages I and II, but at the *coolest* temperature regime during ripening. In other words, the high temperature (35°C) during ripening inhibited, rather than accelerated, sugar accumulation.

In their research on temperature, Hale and Buttrose worked with potted vines in controlled growth chambers. Vine development and grape ripening were fastest when the temperature reached a relatively moderate value, perhaps 27°C–29°C (81°F–84°F). Sometimes experiments with potted plants produce results that are not observed when repeated in the field; however, the McIntyre study discussed prior (among others) was consistent with their results.

Casual observations of early harvests are often attributed to fast ripening. A recent trend to earlier harvests in Australia was thought to be the result of faster ripening caused by decades of regional warming. However, when Victor Sadras and Paul Petrie investigated, they found that the rate of sugar accumulation during ripening had not changed during that time (as shown in fig. 28B).[23] Thus, the maximum rate of ripening may already have been present, and it apparently occurs at a moderate temperature.[24] What *had* changed because of the regional rise in temperature was the earliness of the *onset* of ripening (fig. 28A). The flattening of the dashed curve shown in figure 28A at higher (more recent) temperatures suggests that the latest temperature increase has had less effect than previous temperature increases. This is consistent with previous results—moving from cool to warm conditions accelerates early-season development and leads to earlier bloom (as in fig. 27). This same conclusion—that a significant increase in temperature is correlated with earlier veraison—is drawn by French researchers Eric Duchêne and Christophe Schneider from data dating back to 1972 in Alsace, France.[25]

When vine development has been measured at different temperatures, the experimental observations consistently indicate that the rate of development up to veraison and the rate of ripening have maxima. These limits have gone largely unnoticed, if they have not been completely ignored. In fact, the rate of development leading up to ripening appears to be more sensitive to ambient temperature than the ripening process itself. The early-season conditions that give rise to early budbreak, bloom, and/or veraison may be important in the final grape and wine, even if the rate of ripening remains unaffected.[26]

SUGAR ACCUMULATION VERSUS EVERYTHING ELSE

Anecdotally, many involved in winemaking and wine writing tell us that sugar accumulation does not indicate fruit ripening. Yet, sweetening by sugar accumulation is one of the most fundamental aspects of fruit ripening. The discounting of sugar reflects an experience or attitude that other aspects of ripening—namely, changes in fruity flavors, aromas, and tannin chemistry—are more important than sugar. This approach, however, can be sustained only in warmer winegrowing regions, where getting sufficient Brix each season is almost a given (and is a result of learning how to grow grapes to ripeness!).

The interesting notion that fruit are sometimes ripening too fast is another story in agriculture that is unique to winegrowing. No other fruit bears a similar hypothesis for quality, yet in winegrowing, slow growth is thought to promote "good" and uniform ripening.[27] Rather than hope for ripe fruit at harvest, for vineyards in warm climates, the concern is that sugar accumulation gets ahead of other processes and results in a lesser wine.[28]

The basic idea, which is really quite old,[29] was presented to me by Bordeaux enologist Pascal Ribéreau-Gayon during his visit to UC Davis in 1984. He drew a graph with curves to represent sugar and tannins during fruit ripening, and it showed sugar increasing ahead of the tannins. The graph, he said, illustrated the problem that California faced:

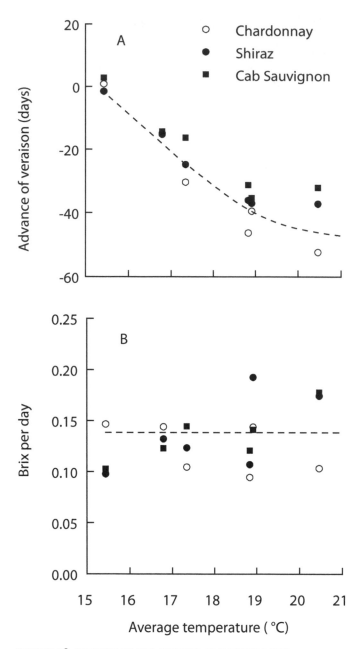

FIGURE 28. ANALYSIS OF THE ADVANCE OF HARVEST DATES
ASSOCIATED WITH CLIMATE WARMING (INDICATED BY AVERAGE
TEMPERATURE DURING THE SEASON) IN AUSTRALIA. (A) Advance
of veraison (beginning of ripening) with increasing average
temperature. (B) Response of the rate of sugar accumulation (Brix
per day) with increasing average temperature. (Redrawn from Sadras
and Petrie 2011.)

it is too warm to grow fine winegrapes in California, because in France the two curves increase together. There are many studies that show "phenolics" increasing during ripening, including some of my early work, but it is unlikely that the measurements used before about 2000 were accurately specific for tannins. At the time of my discussion with Ribéreau-Gayon, it was not understood (as it is today) that tannins were synthesized primarily before veraison (or that the genes for tannin synthesis were essentially turned off at veraison).[30] The perspective that the coordinated increase of tannins and sugar is the proper path to fine wine remains today, despite the knowledge that the synthesis of one precedes accumulation of the other.

There may, however, be other important differences in the temperature sensitivity of ripening processes that could give rise to the overall perception that sugar can increase too fast. Ripening involves the accumulation of many metabolites, as well as the loss of other important flavor compounds such as malate and MIBP (in varieties that make MIBP, that is). Because ripening is defined by sugar accumulation, and the processes that make and degrade flavor compounds all occur during ripening, the increase in Brix is necessarily correlated with other changes in composition that are associated with ripening. This means that Brix, like time, can generally be used to roughly predict progress in other ripening phenomena. The implicit assumption in the "ripening too fast" idea is that sugar accumulation and other processes are only indirectly correlated and are not metabolically linked. Although the discrepancies from 1:1 correlations between sugars and flavors are important for grapes and wines, the coordination (or independence) of various ripening processes with sugar accumulation are not yet understood.

For compounds made before ripening, like tannins, MIBP, and tartrate, there is no metabolic connection, and sugar accumulation cannot progress too fast for processes that are already completed. In the case of red color, imported sugars are themselves well known to be a signal for anthocyanin biosynthesis in plants, and this is true in grapes as well.[31] Thus, there is an established metabolic connection between sugar and anthocyanin accumulation. Experiments at UC Davis have shown that if the import of sugar to clusters is prevented—for example, by girdling (cutting the phloem conduits)—color development stops.[32] That connection would suggest that color should increase with sugar, and it generally does.

It is possible that in warmer temperatures, anthocyanin accumulation "lags" behind the increase of Brix in warmer conditions, depending on the temperature effect on both processes. Sugar accumulation is well known to be relatively insensitive to temperature when compared to anthocyanin synthesis. The temperature optimum for color or tannin synthesis is not well established, but my synthesis of temperature studies from the 1970s up to recently suggests that for Cabernet Sauvignon, color synthesis is maximum near 25°C (77°F).[33] This is consistent with the implicit assumption in the ripening-too-fast concept, although it would be more accurate to describe the situation as I just did (with anthocyanins lagging behind) rather than the received knowledge that sugars increase too fast. (Note too that some varieties, including Cabernet Sauvignon, can produce ample color under warm conditions, although there might be even more development under cooler conditions.)

In other cases, synthesis of important flavor aroma compounds may cease before or continue beyond the end of sugar accumulation. For Gewürztraminer studied in Canada, the rate of Brix accumulation varied significantly among years, but fruit had similar potential aromatic terpene concentrations at common harvest Brix.[34] This is consistent with the results of several earlier studies measuring important aromatic compounds, which showed that the concentration of the aromatic terpene linalool attains a maximum concentration *before* sugar concentration reaches a plateau.[35] No direct metabolic connection is made, but the observations suggest that when conditions lead to more rapid sugar accumulation, there is also more rapid aromatic terpene synthesis. These observations contradict the ripening-too-fast assumption.

It is well known that the rate of loss of malate (usually a bad thing) and of MIBP (usually a good thing in reds today) is temperature dependent, but the temperature responses have not yet been quantified. If we speculate that those processes are more temperature dependent than sugar accumulation, then under warmer conditions, the fruit arrive at a low MIBP at lower Brix (again, a good thing), but also reach a low malate sooner (a bad thing).

The connectedness of sugar accumulation and other ripening processes can be teased out a bit more from some work involving shading. Pruning and shading treatments have been shown to alter the rate of sugar accumulation, but the accumulation of tannins and color did not increase with slower ripening. In one such study, Markus Keller and colleagues imposed shade treatments on whole Cabernet Sauvignon vines for three weeks at the onset of veraison. When the rate of ripening was slowed by shading, so was the rate of accumulation of other phenolic compounds, and the rate of decrease in malate concentration was reduced, which resulted in similar malate concentrations at a common harvest Brix.[36] Grapes with different rates of ripening (measured by Brix accumulation) had similar tartrate, malate, and phenolic concentrations, but slightly different anthocyanin concentrations at a common harvest Brix. These observations show the various ripening processes occurring in conjunction.

In a study that is interesting for its unexpected results, complete light exclusion on Shiraz clusters slowed berry growth in one of three years and did not change Brix slopes (at least to my eye). A different rate of ripening was experienced each year in that 24 Brix was reached four weeks, eight weeks, and six weeks after veraison in 1999, 2000, and 2001, respectively.[37] As in the Cabernet Sauvignon study, fruit with both fast and slow ripening had similar anthocyanins. However, the year with an intermediate rate of ripening had twice the anthocyanins of the other years!

There has been little reference to temperature specifics in this work thus far, and for good reason. There are relatively few studies like the two above with Cabernet Sauvignon and Shiraz in which clear differences in the rate of ripening can be studied for the coordination of ripening processes. The temperature responses of key ripening processes have thus far been poorly studied, and there are few reliable specifics available. This is due in part to the difficulty and expense of conducting controlled temperature experiments. The prudent research in the moderate near term should be directed at sorting out which aspects of ripening are connected metabolically and which can be separated by environment or cultural practices, as the "ripening-too-fast" explanation has thus far found little foundation in empirical evidence.

THE CRITICAL PERIOD: WHY AND WHEREFROM?

As it stands today, "critical ripening" can refer to almost any ripening period—apparently the fact that ripening is occurring is what justifies defining the period as critical. The entire grape ripening period was granted critical status in a 1910 work on Pierce's disease,[38] and nearly 100 years later, the critical period was considered to be similarly long in a report on vineyard mesoclimates in Walla Walla, Washington.[39] If restricted at all, the critical period in winegrowing is described as occurring late or at the end of the season, as in Oz Clarke and Margaret Rand's *Oz Clarke's Grapes and Wines: The Definitive Guide to the World's Great Grapes and the Wines They Make:*

Nevertheless, the weather during the growing season does not make much difference to the quality of the vintage until veraison. Spring frosts or excessive water stress may reduce quantity, but as far as quality is concerned it is the last couple of months before the harvest that really matter.[40]

The later in the season it becomes, the more critical the situation. In *Terroir,* James Wilson quotes Émile Peynaud:

> The saying goes in France, "June makes the wine, August makes the taste." Researchers led by the eminent enologist Emile Peynaud of Bordeaux point out that it is the rate of change of the sugars and acids in the grape during the latter part of the maturation stage that is critical to the strength and quality of the wine. It is particularly those warm, sunny days in September and early October that the winegrower prays for.[41]

In Germany, Canada, and other cool regions, wines have long been characterized by the ripeness of the grapes they are made from. The onset of autumn brings cooler weather and often precipitation, and with that comes the hazard of bunch rots if fruit remain on the vine. There has long been tension surrounding the harvest decision between growers and vintners, or between grower/vintners (winegrowers) and wine merchants. The growers, having no means of warehousing fresh grapes as one can do with wines, shoulder a greater risk of a major loss than a winemaker or merchant when leaving the fruit to hang on the vine. As such, growers favor picking the fruit before fall rains and bunch rots stand to ruin or melt the crop away. In earlier eras, the seller of wines had to deal with wine spoilage, but at that time this was far more fact than risk. Before improved bottling and sanitation, young (rather than aged) wines were sold at a higher price for that particular reason, but today, sick wines are rare.

Historically, a local authority, lord, mayor, or land baron held the power to make the decision of when to harvest. In France, tension between the local authority and the peasant growers was already present in late fourteenth-century Burgundy; it was there that the seigneur authority was held onto the longest.[42]

In Germany, this centralized control is thought to have led to the wonderfully sweet botryticized wine, via the gray mold *Botrytis cinerea* (also known as noble rot). The wine style is attributed to the failure of an order to harvest from Heinrich von Bibra, bishop of Fulda, to reach the peasants on time in the 1775 growing season. Harvest was delayed as the compliant peasants waited and waited to receive permission to pick. The fungal infection set in big time, and the estate owner, knowing that the grapes were no good, left the rotting fruit to the peasants. They made wine anyway, and the sweet taste that resulted was not only good, it became prized and is still intentionally produced in many regions today.[43]

In *Oceans of Wine,* author David Hancock provides a nice description of the conflict between winemaking houses and growers over when to pick in the production of Madeira in the Portuguese Azores, a struggle that dates back to the 1700s.[44] When independent entities are involved, the tension over harvest between growers and vintners derives from the fact that historically, the primary concern for wine quality has been grape ripeness. Later harvests inevitably meant better fruit and thus wine, but picking later also tempts weather-related problems, with previously mentioned fungal rots as the primary concern. For cool climates, where a short season and lack of heat clearly limits fruit ripening, pushing this hazard was and is prudent, because the sugar accumulation and lower acid levels of fruit of a greater ripeness each contribute positively to wine flavors and stability. Obtaining sufficient sugar in the fruit to result in an alcohol concentration high enough to make a stable wine is critical. Recall that Ravaz recommended a Y:PW ratio that produced a wine with less than 10 percent alcohol in the early 1900s, only marginally a wine today. Well known are the practices of adding sugar to the must of grapes that do not reach sufficient sugar concentrations, and of calling the season in which adding sugar was not needed a "vintage year."

The traditional view was, and continues to be, that the grower (who has always had the short end of the economic stick in winegrowing) cannot be trusted to pick the harvest date, because the grower would err on the side of (unripe) caution. This end-of-the-season hazard is diminished in warm, semiarid regions. George Husmann (who contributed to early work on phylloxera resistant rootstock and was an early promoter of Zinfandel in California) described the contrast in 1888, noting how the European winegrowers were challenged each season to ripen their crop, in comparison with the situation in California, where fruit were harvested so ripe that it was difficult to get the fermentation to begin properly.[45] Although the Old World regulatory structure has never been part of New World winegrowing systems, and obtaining ripe fruit remains less of a challenge in warm climates such as those found in California, the decision about when to take (and deliver) the crop has nevertheless remained almost exclusively in the hands of the purchaser/vintner, regardless of location.

Moreover, there is much evidence demonstrating that end-of-the-season activities in the fruit may be a very slow process indeed, as suggested by the curves in figure 2B. The slowing of processes should provide freedom in the harvest decision (weather permitting) and diminish the importance of picking on a specific day, contrary to the concept of a critical ripening period and picking decision. Furthermore, if slow is good, then in climates and seasons with slow ripening, how can the harvest decision still be considered critical? The choice might be hypothetically crucial in bad (hot) climates or poor (hot) years because of rapid ripening, erstwhile running past the optimum point in the season. The decision should be much less of a critical factor, however, in climates and weather that are well suited for growing winegrapes.

THE CRITICAL PERIOD AND PHYSIOLOGICAL MATURITY

"Physiological maturity" is what many winemakers claim to be looking for when evaluating berry ripeness and making the critical harvest decision.[46] Winemakers and journalists report that they are waiting for tannin "ripeness" or "softening" during the latter part of the season.[47] One of the versions of physiological maturity applied to winegrowing involves seed development. As explained by winemaker Christian Moueix, "We used to taste only the juice and skins. Now we taste the seeds as well. When they turn brittle and develop an almond taste, this is an indication that the grapes are *physiologically mature*."[48] The term "physiological maturity" arrived early last century in both plant and human development, where it has relatively specific meanings. With respect to crops, it first appeared in 1906 in reference to seed development in barley. Ironically, the term was introduced to distinguish maturity *from* ripeness, rather than used as a measure *of* ripeness:

> Again, the results would seem to indicate that a distinction must be drawn between physiological maturity and ripeness.[49]

Decades later, the term was invoked in fruit development. The earliest reference to physiological maturity in horticulture appeared in E. V. Miller's 1946 discussion of citrus:

> By *physiological maturity* is meant the age of the fruit with reference to its respiratory climacteric.[50]

Fruit that can complete the ripening process after being disconnected from the parent plant are referred to as climacteric fruit, acknowledging the discovery that the onset of ripening involves a transient and rapid increase, a climax, in the rate of respiration. In this instance, "physiological maturity" was employed to distinguish a measure of fruit development for picking decisions (harvest, after which the fruit could complete ripening) from fruit development to complete ripeness and ready for utilization (e.g., eating or processing). Nearly forty years later, an explicit description of fruit maturity terms was published by the American Society for Horticultural Science (ASHS):

> Physiological maturity—that stage which is: after full development, *before ripening*, and if picked able to continue ripening. This is contrasted with: Horticultural maturity—optimal stage for intended use.[51]

Physiological maturity identifies a stage of development that *precedes* ripening in fruit. Pears, bananas, and so on will ripen (sweeten, soften, and develop color) after being harvested if the harvest occurs close enough to the climacteric period—that is, at physiological maturity. The ripening and sweetening involve a conversion of other carbohydrates in the fruit (usually starch) into sugars.

Grapes are not climacteric—they do not accumulate starch that can be converted into sugars, and they will not complete ripening off of the vine. Thus, winegrape development doesn't lend itself to the usual sense of physiological maturity in fruit. Ironically, the physiological maturity in winegrowing is more similar to "horticultural maturity"—the point at which the crop is ready to eat or process—which distinguishes fruit development from the earlier physiological maturity.

Physiological maturity has been used in seed development research and seed production for decades, usually referring to the maximum dry weight of the seed. This stage in development also often corresponds roughly to the maximum viability of the harvested seed. From the physiologically mature point onward, one could germinate and grow a new corn (or grape) plant from the seed.[52] Grain, however, is harvested long after the maximum dry weight of the seed is attained, as the grain partially desiccates on the plant after its development is complete. Specific definitions for maturity based on seed water content are used for harvesting corn, soybeans, and so on. Thus, for both horticulture fruit and agronomic grain (seed) development, physiological maturity refers to a point in development *before* consumption or utilization. In other words, the point of differentiating between two metrics of maturity was that fresh fruit (and seed) are *not* at their best for harvest at physiological maturity.

Interestingly, the expression for physiological maturity in French (*maturité physiologique*) has a history of its own that dates further back than the use of the term in English. Several nineteenth-century French authors used the phrase in the same sense that I previously described for grain development—as the point in seed development at which the seed will germinate. French botanist Lucien Désiré Joseph Courchet is one example: "For the seed to germinate, it must have acquired what might be called physiological maturity."[53] "Physiological maturity" was also used by early French authors on winegrowing in a manner similar to the horticultural sense in English; that is, to designate a stage of ripening that must occur before fruit can be ready to use or consume: "When at complete physiological maturity, cells of the skin of the berry see their chlorophyll disappear, and they become more transparent in the case of white grapes."[54] This is a description of veraison as physiological maturity, approximately coincident with seed maturity. Similarly, Édouard Robinet (French author of several books on wines and winemaking in the 1800s) distinguished physiological maturity *from* "maturity by convention" where the latter was used for winegrape harvesting.[55]

Despite these precedents for "physiological maturity," winemakers and the wine press have recently adopted the term and use it in a contrary way. References to winegrapes at "full physiological maturity" in the popular press increased in frequency in the 1990s.[56] More often than not, tannins—from seed and/or skin of the grape—were part of the discussion. During the same period, a host of other poorly defined terms entered the popular vernacular, including "ripe tannins," "green tannins," and "phenolic maturity."[57]

The new jargon is associated with real action in the form of later harvests. After decades of harvesting at 21–23 Brix, a move to later harvests grew to become a quality imperative in California at the end of the twentieth century. For example, the harvest Brix for Napa County Cabernet Sauvignon increased rapidly (and as a herd) from about 23 in 1990 to over 25 in the early 2000s.[58] The wines from fruit harvested later are necessarily higher in alcohol, but also typically have deeper color and less astringency and bitterness—in other words, are ready to drink with less aging.

It is probable that a softening of a wine's mouth feel occurs with more mature grapes (despite some evidence to the contrary);[59] however, the nature of this softening is far from understood. According to Karen MacNeil, "Scientists speculate that as grapes reach 'full physiological maturity', small astringent tannins may polymerize, or group together, forming larger molecules."[60] Perhaps a few scientists do speculate; but we are in the midst of a decades-long focus on grape and wine tannins in research and production. It is currently more accurate to say that the term "physiological maturity," as employed in winegrowing, has not been assimilated into the scientific dialogue on grape ripening, and that "physiological maturity," as it has long been used in crop science, has not been assimilated into the world of wine. A credulous few viticulture researchers have adopted the term, but with no clear or consistent use, let alone an explicit definition.[61]

The amount of tannins in red wine closely correlates with observed levels of astringency, but there are other contributors as well. Furthermore, astringency is but one component of the wine mouth feel, which may include some comparison, or balance, with other flavor sensations as we taste. Mouth feel is a complex phenomenon, perhaps so much so that describing it with a single term is misleading.[62] The tannin chemistry is complex, analyses difficult, and progress hard to come by in terms of firm conclusions regarding the taste experience. Some studies do show that the polymerization of tannins (attaching tannin molecules together) progresses with fruit maturity, but the experimental evidence is conflicted, indicating that polymerization leads to both higher and lower astringency. The contribution of seed tannins to wine astringency remains unclear,[63] as most tannin in red wine comes from the skins. Another factor in the mouth feel experience is that as fruit (including winegrapes) ripen, the cell walls undergo disassembly and pectins are released. These pectins may have a sensory effect of their own that softens mouth feel,[64] and the cell wall fractions may also interact with the tannins, reducing tannin concentrations or otherwise altering the finished wine. Before the recent movement toward later harvests, there were few studies conducted connecting late harvest fruit maturity and wines. Thus, the academic world must now play catch-up, in order to evaluate fruit that is harvested later in the season, as well as the resultant wines.

It is interesting that whereas the viticulture academics may have gone overboard in relying on their measurements of Y:PW and GDD to define the conditions for good grapes, winemakers have moved away from objective inputs (such as Brix) for assessing fruit development, preferring to declare maturity based on the subjective taste experience of the winemaker (although there are some attempts to formalize taste analysis). The taste of the fruit may well be predictive of the wine, but in contrast to the long history in crop science where "physiological maturity" has definition(s) that are both objective and explicit, the use of the term in the world of wine is vague and approaches a subjective "when I say it is best" argument. Furthermore, as employed in the harvest decision, "physiological maturity" is the virtual opposite of the explicit definition and broad use of the term in horticulture. It smacks of what lexicographer H. W. Fowler calls "vogue words," showing "the delight of the ignorant in catching up a word that has puzzled them . . . and exhibiting their acquaintance with it as often as possible."[65] However, the use of physiological maturity by winemakers may be similar to the use of vine balance in the vineyard, in that both involve an expert's integration of inputs into a judgment of correctness.

THE CRITICAL RIPENING PERIOD:
EARLY IS IMPORTANT, TOO

In contrast to the traditional emphasis on the end of the season, important aspects of fruit development early in the season and the sensitivity of the final fruit composition to early-season conditions have been largely overlooked. When important compounds such as tartrate, tannins, and MIBP are synthesized only early in the season—before the onset of ripening—it is an inescapable truth that early-season fruit development is important too. Although the synthesis of tartrate shows little environmental sensitivity, the concentrations of tannins and MIBP are more responsive to conditions before ripening than during.[66] Late-season leaf thinning to avoid veggy wines has been replaced by earlier canopy management because it was found that the veggy MIBP is greatest when the berry is heavily shaded early in the season; late-season light has limited effect on its final concentration in the harvested fruit.[67]

Of course, many taste and aroma compounds that are important in wine flavor accumulate from veraison onward, but surprisingly, pre-veraison conditions are also important in determining the concentration of anthocyanins that are made after veraison. In several studies, water deficits and shade have had as great or even greater effects on fruit growth and composition, as well as final color in red grapes, when imposed before veraison, rather than during ripening.[68] The earlier in the season that shoot-hedging or cluster-thinning practices are imposed, the greater the response that is usually observed.

It is not accurate to assume that the concentrations of flavor and aroma compounds increase post-veraison, and that the later the harvest, the better. There are hundreds of aroma compounds, and several show no clear increase, or even exhibit a decrease during ripening. In a study that extracted volatiles from Cabernet Sauvignon and Riesling grapes throughout development, "both Riesling and Cabernet Sauvignon grapes had a more complex volatile compound composition pre-veraison than post-veraison. This study suggests that some compounds that contribute to grape aroma may be produced pre-veraison, and [do] not simply accumulate after veraison."[69] The concentrations of different compounds naturally peak at different times during development, and in some cases that timing is affected by environmental stress. This means that there is unlikely to be a single optimum time; rather, there is a need for judgment in harvest decisions based on stylistic objectives. There is much to be learned about flavor metabolism in grapes.

STRESSED VINES MAKE THE BEST WINES

The notion that the best wine results from stressed vines appears in wine journalism, travel guides, wineshop blogs, and other popular press venues. According to a 2009 *Business Week* magazine article,

> If you're a devotee of small producers or high-end wines of any kind, most likely you've heard the phrases "reduced yields," "dry farming," "nutrient-poor soils," "high vine density," and more. These practices are regularly employed by many of the world's best winemakers, and they all have a single goal in common: to stress the vine. It is now common knowledge (and common practice) that vines pushed to the edge of their tolerance for many environmental factors generally tend to make better wine—more concentrated, more complex, more tasty.[70]

A severely stressed vineyard would not make or ripen fruit, and pushing grapevines to the edge of their tolerance would be tantamount to driving oneself out of the vineyard business. There is, however, truth to be found in the stressed vine belief, and the following section will attempt to briefly journey through the fact and fiction of stress in winegrowing.

In this concept, "stress" refers to environmental perturbations—such as temperatures or water status—that inhibit growth. The stressed vine belief is the antithesis of the contented cow notion, for which there is good evidence that high temperature stress reduces productivity and quality.[71] As with cows and milk, the best winegrapes are not thought to come from extreme environments—the hottest, coldest, wettest, or driest places where grapevines can be grown—nor do they hail from the places with the most or least sunlight, or with soil containing the most or least amount of a key plant nutrient. Thus, the stressed vine maxim is incorrect in a broad, general sense.

Stress is usually defined as an environmental condition that impairs plant growth. Stresses can be biotic (e.g., diseases) or abiotic (e.g., freezing temperatures). Plant growth is sought in most crops, but excess vegetative growth is a common concern in winegrowing. Excess shoot growth can compete with fruit and buds for resources, resulting in poor microclimates for both grapes and the buds that will be next year's grapes but good microclimates for fungal diseases. Growth in plants is most sensitive to water deficits compared to perturbations in other environmental parameters such as temperature, light, or nutrient status. Mild deficiencies in water and some nutrients, especially nitrogen, can be beneficial in that they reduce the otherwise excess vegetative growth. Because vegetative growth is more sensitive than cluster initiation or other aspects of fruit development, the right conditions accelerate fruit ripening and even improve it. The clearest case of beneficial stress, with respect to fruit and wine composition, is when red winegrapes experience moderate water deficits and the fruit respond with greater tannins and color.

Vine Water Status

As noted in chapter 1, in the discussion of myths related to yield and berry size, part of the stressed vine concept derives from the historical fact that the vineyards giving rise to premier wines were not irrigated. Not only have European vines generally been dry-farmed (without irrigation), but irrigation was widely banned in Europe, ostensibly to control quality. Even in southern France, where classed growths were less developed and where canals were constructed in the mid-1800s to deliver irrigation water to fields, irrigation was still not used on vineyards.[72] There may have been some practical experience driving the aversion to irrigation, but a focus on irrigation per se arises from a lack of understanding of how plants and grapevines use and respond to water.

Grapevines are typical plants in that they use a lot of water in the course of living and growing. Plants use more water in warm, semiarid climates, which often do not supply the water necessary to grow and produce crops. If it is not available from water stored in the soil or delivered by rain, water must be supplied by irrigation or growers must forgo production. Irrigation is not bad for winegrapes in and of itself; it is water, something a plant needs to survive, grow, and ripen fruit. Grapevines have no sense of how water arrives to the root surface where it is taken up. As Maynard Amerine of UC Davis wrote more than forty years ago, "Since moisture is moisture to the vine root, the opinion that irrigation is detrimental to wine quality compared to rainfall is erroneous. . . . If irrigation were injurious, most of the wines of Europe would be poor, for summer rainfall occurs to some extent throughout the northern European vineyard areas."[73]

Plants experience water deficits to varying degrees, depending on the availability of water to the roots and the demand for water from the dry atmosphere. Because vines experience a water status (a relative wetness or dryness) that greatly affects growth and ripening (regardless of whether they are irrigated or not and whether it rains or not), it is not the rain or irrigation that matters to the plant but rather how wet or dry the plant is. In much of California and other places with Mediterranean climates, it typically does not rain significantly for four or five months during the growing season. This can be a good thing, diminishing pest and disease pressure, and facilitating grower control over vine water status. But the vines (indeed all plants) still need water to keep stomates on the leaves open to conduct sufficient photosynthesis in order to grow, ripen fruit, and nourish buds for the next season.

Fortunately, the myth that fine wine cannot be made from irrigated vineyards is fading. As the grapevine's real-world need for water has been slowly acknowledged, restrictions on irrigation have been rolled back. Irrigation is now allowed in most of Europe, although it remains essentially illegal in appellation control areas without special permission.[74] The need for irrigation in some climates has also been assimilated by part of the wine press, including in the works and opinions of Hugh Johnson[75] and Robert Parker.[76]

Where irrigation has been essential for the viability of vines and to produce a crop, irrigation has been allowed. Irrigation and public water projects helped create the West as we know it in the United States, and water has been important in California grape production from the outset in the nineteenth century. Anaheim, east of Los Angeles, was an important viticulture area in the mid-1800s, and the rise of orchards and vineyards there was completely dependent upon the development of an irrigation system.[77] There was early advice against irrigation (usually with the admonition that irrigation reduces quality), but it could not be heeded, nor was it.[78] Water has been an important factor for another New World upstart, Australia, in the twentieth century.[79] Faced with the obvious need for irrigation, Australian wine journalists were never on board with the myth of irrigation as a wine spoiler. Although most of the Old World winegrowing traditions were emulated in

California winegrowing (as much of the winegrowing was the work of European immigrants), the traditional attitude toward the outlawing of irrigation in winegrowing was not feasible in much of the state—and the same goes for most of Australia's winegrowing regions as well.

Many vineyards located in Californian winegrowing regions that are well known for high wine quality are irrigated today, but were not irrigated in the past. There was no irrigation whatsoever in California's North Coast counties such as Napa or Sonoma prior to 1900; even into the 1970s, irrigation was often used only in the first few years to help vines become established. Yet, as irrigation has become widespread, there has been no call to action because of the decline of the wines. The old myth that irrigation is bad is giving way to more refined knowledge of how the vine responds to more or less water. It is the vine water status that affects vine and berry growth and metabolism, not the amount of applied or precipitated water.

Vine water status is a consequence of the soil water supply and aerial demand for water. Research into water management in vineyards has been active in semiarid California and Australia, and from such research has emerged a clearer (yet still developing) understanding of grapevine water relations. Nevertheless, where irrigation is needed, there remains the challenge of supplying the correct amount at the correct times.

Winegrapes ripen faster or more slowly depending on vine water status. When water deficits have been imposed by withholding irrigation, sugar accumulation has increased or remained unchanged under mild stress, and may decrease with severe stress. Studies in both California and Australia have shown that moderate water deficits usually result in earlier sugar and lower acid.[80] In most cases, fruit color and tannins increase in response to water deficits, as water deficits produce a double whammy of smaller berries and increased color synthesis—changes in grape composition and wine sensory attributes that are often perceived as positive. Water deficits make smaller berries in general, although berry growth is not as sensitive to water deficits as vegetative growth.[81] Wine writer Karen MacNeil recommends avoiding irrigation at the end of the season, but does not comment on the early-season role of deficits in promoting tannin synthesis, nor does she acknowledge the experiments of Bryan Coombe that showed no response of berries to late-season irrigation.[82] Robert Parker has referred several times to the dilution of flavor by late-season rains;[83] however, berries are covered in wax, and laboratory experiments show that when berries are completely submerged in water for forty-eight hours, the volume increases only about 4 to 5 percent.[84] The water uptake for berries in the field, which are not submerged and have part of their surface protected by contact with other berries, must be much less.

Water deficits inhibit shoot growth even more than fruit growth. This can be positive, but like all of these stress responses, it depends on how much inhibition takes place. Reducing shoot growth can reduce the cluster shading that promotes higher concentrations of veggy MIBP. On the other hand, clusters exposed to direct sunlight in hot conditions can become shriveled and sunburned. Severe water deficits can fail to ripen the crop because photosynthesis is drastically inhibited or the leaves are lost altogether. Overall, a rough generalization is that some water deficit is beneficial, especially for red varieties; the sensory impact of water deficits in white winegrape production is not clear. In either case, if the climate is dry and/or the soil shallow, vine health is risked when water deficits are imposed.

Sensory difference tests conducted at UC Davis showed that wines made from vines exposed to early deficit (water restricted early in the season), late deficit (water restricted late in the season), and no deficit (continually well watered) treatments differ from one another.[85] Thus, in addition to water deficits per se, the timing of the water deficit is of practical importance to wine flavor and appearance. This early work and the few other studies like it, however, didn't go very far toward resolving the consequences of water deficits, yield, or berry size for wine flavor and quality. In fact, few irrigation studies are carried through to sensory analysis (as shown in table 4).

Mineral Nutrition

In addition to water stress, grapevines are subject to stress from deficiencies and toxicities of mineral nutrients. There are no benefits of toxicities, but the consequences of nutrient deficits for wines depend somewhat on how deficiency is defined. In the conventional sense, deficits are defined by reducing yield. Standards for adequate soil or grapevine nutrient status are determined based on the concentrations required to attain maximum yield, but high-end winegrowers are more concerned about fruit quality. Standards for the nutrient statuses that result in optimum winegrape composition have not yet been established. In fact, there are few clear and specific consequences of mineral nutrition for fruit composition. On the other hand, for most nutrients, there are severe consequences for growth, flower development, fruit set, and ripening when a nutrient is clearly deficient.

There is some evidence that low nitrogen inhibits the synthesis of aroma compounds, and yet promotes color in reds.[86] High soil fertility, especially high nitrogen, may push the vine toward the unbalanced vegetative/reproductive growth of the cultivated vine in the manner described by Bioletti (discussed in chapter 2), in which fruitfulness is diminished in favor of vegetative growth. Very low nitrogen in fruit can promote synthesis of bad-smelling hydrogen sulfide during fermentation, and very high nitrogen can lead to the production of undesirable urethane in wines. High potassium availability in the soil has been associated with high pH in musts and wines, but this phenomenon is poorly documented. Consequently, Markus Keller at Washington State University proposed the "regulated deficit nutrition" concept in an analogy to regulated deficit irrigation,[87] a term often used to describe the imposition of water deficits to improve red winegrapes.

The most common result of a nutrient deficiency is an inhibition of photosynthesis and acceleration of leaf senescence that reduce all other growth and development, and can slow or prevent complete fruit ripening. While there are problems that arise with nutrient excesses, managing nutrients for sufficiency rather than stress (deficits) is prudent viticulture.

Fruit Exposure and Light-Temperature Interactions

An important cultural practice in winegrowing is "canopy management"—the manipulation of shoots and leaves in order to distribute leaves and fruit in accordance with production objectives. Although objectives involve air flow, most attention is given to the distribution of light inside the leaf canopy. In the 1980s and 1990s, Richard Smart helped draw attention to the cluster and bud microclimate in both academic research and production viticulture. For example, Smart's third principle of canopy management states: "Canopy shade should be avoided, especially for the cluster/renewal zone."[88] It became a common production objective to have a more open canopy and more fruit "exposure." But it is not correct to extrapolate the idea that if a little more light is good, a lot more light is best. The initiation of grape clusters and the synthesis of anthocyanins and other phenolics (tannin-like molecules) are promoted by light on the buds and fruit, respectively, but this occurs only up to a relatively low light intensity. In one of the best studies thus far of the cluster light environment, berry growth and composition in both Cabernet Sauvignon and Pinot noir improved as light on the fruit increased to about 10 percent of full sunlight; however, light intensities greater than 10 percent of full sunlight had no further beneficial effect.[89] Similarly, reducing cluster shade was effective in reducing the veggy compound MIBP but only for light intensities up to about one-third of full sunlight.[90] Thus, the available evidence indicates that a low amount of light on the fruit is sufficient to realize the benefits of light in ripening, beyond which berries may overheat.

Just as in the supply of water or heat, there is a limit for light exposure, beyond which it can be harmful, particularly in warm climates. Light is energy that heats up berries, which may be good in cool conditions, contributing, for example, to the loss of MIBP. High berry temperatures, however, also cause the loss of desirables such as malate and red color. In some climates, berries can get sunburned, a problem experienced on hot days in vineyards that were taken to very open canopies, in accordance with the "if a little cluster exposure is good, more is better" idea.

We need to know the light and temperature optima for important processes like the synthesis of tannins, anthocyanins, fruit esters, and the aromatic terpenes in order to design training systems and canopy management to result in the best cluster microclimates. We already know that varieties differ in their light responses and probably in their temperature responses as well, so we can expect to have different objectives for different varieties. The simple "more light is better" approach was taken as the Vertical Shoot Positioning system was widely adopted. This system supports the shoots so that clusters are often exposed to full sunlight. The experience in some high light and temperature climates showed that grapes will be damaged when limits are exceeded. In order to establish valid canopy management objectives, we need quantitative answers to the questions of light and temperature effects on the fruit.

Acclimation

The stressed vine model is sometimes described as akin to "better grapes through strength training."[91] One way in which this analogy is thought to operate is the idea that wide spacing detracts from quality (which we reviewed in the HYLQ chapter), because the vine roots tend to spread lazily horizontally, rather than working down through the soil layers.[92] "Working" and "lazily" are human characteristics, and the latter is an attitude toward the former. Analogy can be helpful in conveying concepts, but this language takes it too far and obscures how plants truly function—without attitudes.

Nevertheless, exposure to some stress conditions results in increased resilience in subsequent exposures. This phenomenon is called acclimation, and it bears some analogy to training. Leaves that have experienced water deficits can sustain photosynthesis to drier conditions than leaves without that prior experience,[93] and grapevine leaves also exhibit acclimation to light[94] and temperature,[95] in terms of increased resilience of photosynthesis during stress. Acclimation is a physiological adaptation to the environment that probably helps the vine continue to deliver sugar and amino acids to ripening berries during stress, and in that way could contribute to better fruit than would be possible without acclimation. But the training idea and acclimation in plants are means of ameliorating stress, rather than enhancing responses to stress in the hope of obtaining better fruit.

CONCLUSIONS

A ripening period that is truly critical would be a narrow period of time during which important and rapid changes in fruit composition would naturally occur. These rapid changes would render the timing of the harvest decision critical, in order to capture a fleeting optimum fruit composition. Physiologically speaking, things are actually changing slowly at the end of the season. Accordingly, no critical period has been identified, except in the case of impending fungal outbreaks or other weather-related disasters. This aspect is trivial, in that it is indeed critical to have a crop to harvest, but the key decision-making aspect is one of predicting the weather, not assessing grape quality or maturity.

A rapid rate of ripening would contribute to a narrower window of optimum fruit maturity and the hazard of missing that optimum. However, many viticulturists and others in winegrowing have mistakenly assumed that warmer conditions inevitably lead to faster development and ripening (and that growing degree days can predict fruit maturity). While this can be true in low temperature climates, the foregoing analysis of research paints a different picture, with a moderate temperature maximum for rates of vine development and sugar accumulation, and greater impacts on preripening development than on the rate of ripening. Embedded in the ripening-too-fast-myth is another assumption, that when ripening occurs under warmer conditions, the rate of sugar accumulation increases but other flavor metabolism does not, leading to imbalances of fruit composition. Aside from the previously mentioned insensitivity of sugar accumulation to high temperatures, the evidence we reviewed regarding imbalances based on different rates of sugar accumulation was inconclusive. However, it is likely that various aspects of berry growth and ripening have different temperature responses, and that since these processes take place in the berry (i.e., at the same temperature), those differences are inescapable, unless they are separated in time.

Alternatively, a critical period could also be a narrow period of unusually high sensitivity to environmental conditions at any time in the season. Yet, flavor molecules (both desirable and undesirable) are synthesized both during and before ripening. Early- and late-season light, temperature, and water status conditions are important in determining the final composition of the winegrape. Thus, fruit ripening responds to cultural practices and environmental conditions throughout development. The potential impact of the environment on fruit composition diminishes as harvest approaches and more of the season is "over with." Indeed, aside from acute problems with heat spells or disease outbreaks near harvest, most environmental impact on harvested fruit composition occurs well before harvest. Focusing on the harvest decision at the end of the season prevents winegrowers from operating with (or discovering) a complete understanding of the conditions that occur throughout the season that affect final winegrape and wine composition.

Furthermore, if the harvest date is indeed critical to wine quality, mistakes of picking too early and too late would sometimes be unavoidable, commonplace, and apparent. Where are these anecdotes or wine reviews? Has this critical harvest decision never been made incorrectly? Similarly, we should also expect that the investment in getting the crop off the vines quickly and processed would be much greater than it has been. Top-end producers would crow about their investments in winery capacity, in order to allow for processing all fruit at the critical harvest time,[96] and winemakers and the wine press would bemoan the failure to invest in facilities that can exploit this "single most important viticulture decision."

Perhaps we don't see these scenarios playing out because, although wine characteristics depend on Brix at harvest, the differences from the choice of harvest date have more to do with wine style. Of course, it is important to harvest ripe fruit—neither green and harsh nor oxidized raisins, but highly respected wines have been made and appreciated when coming from grapes at a wide range of sugar concentrations—

that is, harvest time decisions. For example, only 8–10 percent alcohol was considered adequate in the nineteenth century (an 1863 Haut-Brion sold in London had 11.5 percent alcohol, which was high for the period),[97] about 12–12.5 percent became the norm in the twentieth century, and now more than 15 percent is commonplace in high-end reds.[98]

Later harvesting has led to controversy, as some refer to later harvest wines as "fruit bombs" and to the earlier harvest products as "finesse wines." Old World styling is associated with lower Brix and "finesse" wines, which the New World makers initially sought to emulate. The higher Brix is more accessible in warmer climates such as those of California and Australia, and that contributes to the Old World–New World tension, but it is also the natural progression of the simple quality = concentration idea that drives the HYLQ and BBB myths.

Despite progress in viticulture and the increasing average Brix at harvest, the weather is still a factor that can dictate ripeness. This can be considered quality or style, perhaps depending on how low the Brix is at harvest. Climate change, and the warming in northern Europe in particular, has created more flexibility for those producers to produce later harvest wines. European winegrowing regions that were the source of the HYLQ concept (and where the climate was marginal for ripening grapes) increasingly allow winegrowers there the opportunity to produce wines from riper fruit if they so choose. With Old World and New World each presented with increasingly similar opportunities to select the fruit maturity at which to harvest, the resultant wines in the future may be accepted more as styles, and less as the objects of an argument regarding right and wrong. These stylistic issues stand in contrast to the simplified version of fruit and wine quality that depends on concentration alone.

The 2011 season in Northern California was extraordinary in that it was unusually late and cool. This left many harvesting fruit at Brix lower than had been assumed was necessary to create the best wine. Some reviews reported no fruitiness and overpowering vegginess in Cabernets with "moderate" alcohols of 13.5–15 percent; however, the reaction from others was not that wine would be lesser; rather it would have more Old World "finesse." As wine writer Jim Boswell said, "Alcohol levels are lower in these wines, and the best ones will feature elegance and finesse. These are not qualities normally associated with brawny Napa Valley Cabernet Sauvignon."[99] In this case for this wine expert, it was only a matter of style. And, indeed, some feel strongly that the earlier harvest is better.

Some winegrowers and critics have steadfastly resisted the call for physiological maturity and late harvests, including Randy Dunn, a producer of fine Napa wine, who referred to waiting later and later and looking at the seed to determine fruit maturity as "bullshit."[100] Dunn apparently prefers the traditional Old World model. I recently visited a former student who grows grapes and makes them to wine in Napa Valley, and tasted one of his (relatively) early harvested Cabernet Sauvignon wines. For me, it was wonderful—and reminiscent of fine Napa wines from the 1970s and 1980s. The producer said he thinks that earlier harvesting catches more red and fresh fruit flavor, and our recent research agrees.[101] Regardless of subjective opinions about the late harvested and heavier wines, wines of earlier decades are not held in disrepute because they were made from grapes harvested too early.

Whether winemakers, wine critics, and consumers fall on one side or the other regarding later harvested wines, the controversy generates heat and takes us toward people's preferences and away from the grapevine. There is even more heat generated with terroir, the biggest myth of all, which we will strive to connect to the grapevine in the next chapter.

CHAPTER 4

The Terroir Explanation

A belief is not true because it is useful.
—Henri-Frédéric Amiel

Terroir is a ubiquitous term in the commercial wine world. I can't visit a winery's tasting room or website and seldom make it through a dinner party without having it thrust at me, often it seems in order to test my wine savvy. If you are even just a casual appreciator of wine, chances are you have encountered a situation in which you were called upon to declare whether or not you were "in the know" with respect to terroir. At some point, earlier in your wine drinking career or perhaps just now, the question arises: what does *terroir* really mean?

Terroir means "soil" or "land" in straightforward French, but when used in the wine world, things get a bit more hairy.[1] It was employed in reference to wine initially in the phrase *goût de terroir,* where the phrase translates to "the taste of the soil," and folks often meant what they said quite literally. It has now come to encompass the taste of a place, such as the taste of a wine that is associated with a place. *Terroir* is a loanword in English, again used initially as *goût de terroir,* or "taste of terroir," wherein it is for some evidently possible to translate *goût* but not *terroir*. The word has been and is still used today as an explanation for what makes good winegrapes, and hence, fine wines. Many wine journalists, winemakers, and industry influencers claim to be engaged in a never-ending search for a true or natural expression of terroir in wine, but what does that mean?

According to *Newsweek* magazine, "Global wine tastes may be shifting toward big, bold and fruity, but the high end of the wine market is still dominated by the regions of France, where nuance and complexity are inextricably bound up in a region's traditional grapes, the unique qualities of the soil and the local weather—what the French call terroir."[2] Beverage consultant, vineyard owner, and wine producer David Skalli asserted that 99 out of the top 100 wines are there because of terroir; top producers "have nearly perfect fruit, climate, and terroir." Bruno Prats (wine producer, and at the time owner of Bordeaux winery Château Cos d'Estournel) stated:

> The very French notion of terroir looks at all the natural conditions which influence the biology of the vine and thus the composition of the grape itself. The terroir is the coming together of the climate, the soil, and the landscape. It is the combination of an infinite number of factors, to name but a few. All those factors react with each other to form, in each part of the vineyard, what French wine growers call a terroir.[3]

Matt Kramer, cooking and wine author and columnist for *Wine Spectator*, writes that terroir

> is everything that contributes to the distinction of a vineyard plot. . . . But terroir holds yet another dimension: It sanctions what cannot be measured, yet still located and savored. Terroir prospects for differences. In this it is at odds with science, which demands proof by replication rather than in a shining uniqueness.[4]

What did we just learn about what makes a good grape and thus a good wine? As Mr. Prats notes, the composition of the grape itself is foremost in any effort to gain insight into the vineyard side of winegrowing. This is consistent with another mantra of winegrowing: that wine is made in the vineyard (as opposed to the winery). No doubt what happens in the vineyard is important, although one might suspect that some of what happens in the winery is important as well. Our sources tell us good wine is due to terroir, that terroir is related to the environment, and also that terroir is more than soil. There is something in there about uniqueness; then the explanation of terroir becomes blurry, "bound up in it," instead of clear. According to Prats, terroir has lots of parts, even an infinite number.

Kramer seems to warn that an empirical or objective analysis regarding terroir is futile. This appeal to the complexity of terroir is sometimes linked to claims of the inexplicable, ineffable, and/or mystical. Bernard Ginestet (author of several short books on various French winegrowing areas) told writer William Langewiesche that he could spend days trying to understand terroir.[5] I've spent years.

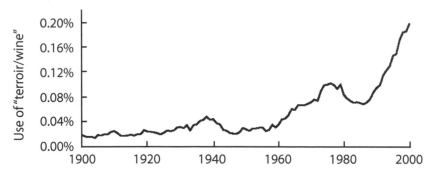

FIGURE 29. GOOGLE NGRAM ANALYSIS OF OCCURRENCES OF "TERROIR" RELATIVE TO "WINE" IN DIGITIZED BOOKS PUBLISHED IN ENGLISH FROM 1900 TO 2000. Despite being associated with wine since at least the seventeenth century, use of *terroir* in winegrowing has experienced a recent and dramatic increase.

Using the term *terroir* with respect to wine in English is a surprisingly recent phenomenon. In an analysis of all the digitized books published since 1900, a Google Ngram shows that *terroir* appeared in conjunction with wine infrequently until the 1980s, at which point a sharp increase began that continues to this day (fig. 29).[6] The timing of the transition is taken up later in this chapter, but for now it should be noted that there was no discovery in the vineyard that led to a revelation regarding the role of terroir in winegrowing.

ANCIENT ORIGINS OF THE TERROIR CONCEPT

Any understanding of the place of terroir in relation to wine quality requires a look back. Terroir is clearly an old concept. Old World supporters of the terroir explanation argue that it is its age that in part validates the notion, and their long history with terroir (they have been working at it for hundreds if not a thousand years) establishes their high standing in the wine hierarchy. But when does the terroir concept truly begin?

For some authors, terroir began almost as soon as winemaking began. Italian wine producer Mario Falcetti argued (in his opinion paper "Terroir: What Is 'Terroir'? Why Study It? Why Teach It?") that the ancient Egyptians (ca. 3000 BC) understood terroir, and that its importance was well recognized in Roman viticulture, as indicated by the writings of the Georgians (ca. 200 BC–AD 200).[7] Falcetti's criterion seems to be that terroir is understood when it is reported that some wines are better than others depending on the region or perhaps the vineyard source. In this use terroir is little more than an acknowledgment that wines are different, but hardly an explanation of *why* wines are different. In order to gain insight into how to go about growing a good or better wine, we need to know whether it is the place itself that is key (as opposed to the people or their practices at the place), and if so, how a place matters and contributes to final wine quality.

We know that wine is derived from the fruit of grapevines through the yeast-driven fermentation of the sugars found in ripe berries, and (usually) the bacteria-driven fermentation of malate, an organic acid. During yeast fermentation, sugars are converted to alcohol, and other compounds in the grapes are modified to become volatile aromas. Additional malolactic fermentation by bacteria produces lactic acid (which is less sour than malic acid) and diacetyl (which has a buttery aroma). For red wines, it has been fairly well established and generally accepted that most flavors and aromas originate with the fruit. For white wines, there is more of a contribution to aroma and flavor from the yeast metabolism compared to reds, but this may also depend on fruit composition. Recently improved methods for studying the hundreds of taste and aroma compounds have helped reveal that there are more yeast-synthesized flavor compounds than previously thought. In terms of the grapevine itself, wines differ in flavor attributes because the fruit composition (the absolute and relative amounts of flavor compounds) differs.

Terroir is applied to explain all those differences in the fruit. But what could have been understood about what makes wines different when the terroir explanation developed hundreds of years ago? From our earliest written records, plants (and therefore grapes) were once understood to be comprised, quite literally, of the soil in which they grew. Reaching back to the ancient Greeks, we find that Empedocles (490–430 BC) held that all things were comprised of the four elements: earth, water, fire, and air. Hippocrates (460–370 BC) connected the four elements to four humors: blood, yellow bile, phlegm, and black bile—each fluid was considered to be derived from a specific organ. A theory of physiology and disease called humorism developed from these connections, in which the state of one's health—and by extension, the state of one's mind or character—depended upon a balance among the four elemental fluids. Aristotle (384–322 BC) incorporated the four elements and humors into his description of the natural world, including plants.

The four elements were mutually convertible into one another, and in various combinations were believed to form all of the objects in the natural world. With only four elements, there were limited options for considering the source of grapes and wines. Aristotle claimed that plant roots engulfed minute particles in the soil, which were preformed miniatures of that particular kind of plant. According to Aristotle, the roots acted like mouths in eating, and there were specific soils that corresponded with each plant that were correct for consumption. "The roots are the superior part of a plant, for from them the nutriment is distributed to the growing members, and a plant takes it with its roots as an animal does with its mouth."[8]

Humorism, or the doctrine of the four humors, became the predominant medical theory and retained its popularity for centuries, largely through the influence of the writings of Galen (AD 131–201). Galen extended the animal analogy of plant growth and described the earth as a stomach for plants:

> For animals, inasmuch as they are not made to grow by the earth, apart admittedly from a few instances, Nature has fashioned the stomach as a storehouse of nutriment, like the earth for plants. In plants, then, the earth serves in the manner of the stomach, continually irrigating a readily available and bounteous nutriment so long as the seasons be in accord with nature from Zeus.[9]

> For just as plants draw all their nourishment from the earth through their roots, so the heart draws air from the lungs.[10]

Galen's Aristotelian ideas on medicine, which made extensive use of wines as treatments for wounds and illnesses, held sway for about 1,500 years, until challenged by Paracelsus (1493–1541) around the same time Copernicus was questioning the Earth-centric view of the cosmos. Paracelsus was an iconoclast in his own way, but still given to an intuitive "because it seems to make sense" doctrine. For example, the word *orchid* derives from the Greek *orchis,* meaning "like a testicle," because of the plant's ball-shaped double root. To Paracelsus, this resemblance meant that the plant was capable of curing venereal diseases.

Such was the thinking of the time—that grapevines were literally of the soil on which they grew. Of course, humorism (and much of the ancient understanding of nature) incorporated myths that have since been set aside by science, but often not quickly, and in some cases not readily. Sherwin Nuland in *The Mysteries Within* describes how Galen developed the mythology surrounding our understanding of the human body and its physiology into medical principles that had remarkable resilience.[11] Although there is less of a mythical record on plants and viticulture than on medicine, a parallel phenomenon has been observed in which ideas about plants and their function derived by analogy (which were decidedly off the mark with respect to reality) have long persisted, even in the face of clear contradictory evidence.

One example of this reasoning by analogy with respect to plants is provided by cotton. The Greek historian Herodotus (484–425 BC) wrote about a tree in Asia that bore cotton "exceeding in goodness and beauty the wool of any sheep." Cotton seemed similar to wool, and it was reasoned that it arose as wool did—thus, the vegetable lamb myth. As recounted by Henry Lee in *The Vegetable Lamb*, one version appeared in *The Travels of Sir John Mandeville*, which explained that Mandeville, upon returning from India in 1371, claimed to have found a wonderful tree which bore tiny lambs on the ends of its branches. These branches were so pliable that they bent down to allow the lambs to feed (on the grass) when hungry.[12] Mandeville's account of the vegetable lamb (fig. 30A) spread through England. Although in 1641, Kircher of Avignon described cotton and declared it to be a plant, the idea of the vegetable lamb was still worthy of scientific comment in 1725, when a German doctor named Breyn (in a communication to the Royal Society of London) emphatically stated that the story of the vegetable lamb was nothing more than a myth.[13]

Another example of this kind of reasoning is the mandrake. Once called by Shakespeare "the insane root that takes reason prisoner," the mandrake has been given many attributes based on its sometimes human-like appearance (fig. 30B). It was carried by sailors and hung over doorways for protection, and it was believed that sleeping with it under a pillow increased fertility. The mandrake even made an appearance in the Harry Potter movies, which highlighted the mythical hazards of uprooting a mandrake plant without following careful instructions. In addition, botanist and historian Conway Zirkle describes tales from medieval times of imaginary hybrids created by grafting one plant onto another.[14]

Similarly, the ancients give us one origin of the terroir concept— flavors of good wines come from vines eating the right kind(s) of soil. The concept of "organic" nutrition of plants, in which the living plant takes up food (even living stuff) for nourishment, held for more than 2,000 years, until the mid-1800s. Whether terroir is legend or reality is an important question for those interested in improving winegrowing.

FIGURE 30. VEGETABLE LAMB AND MANDRAKE. (A) Two versions of the mythical "vegetable lamb" plant that produced cotton. (Image from Lee 1887.) (B) Illustration of the mandrake root, which sometimes bears resemblance to the human figure, and has been assigned many mythical properties, evidently on that basis alone. (Image from *Tacuinum Sanitatis* [1474], a medieval handbook on health and well-being.)

TERROIR ORIGINS IN MEDIEVAL EUROPE

There are two stories about the basis for terroir that are somewhat at odds with one another: one that is presented in the popular wine press, and another that has developed through scholarship in geography, history, and the social sciences. The popular notion continues with the soil-based appellations in wine, whereas the alternative story is more of a social construct for establishing (and maintaining) control of both territory and wine reputations.

Wine regions, wineries, and wine consumers have all developed an affinity for the past in the wines they produce, sell, and consume. The first appellation, for instance, has several claimants. According to Marcel Lachiver's history of French winegrowing, an appellation was established and in operation in Paris in the twelfth century: "The *appellation* dates from 1175. Nevertheless, it isn't until 1416 in a policy order that a specific border [bordered area] is defined as the source of an appellation, but the text does nothing but validate this practice."[15] Claims of the first appellation come from several places outside of France in the early eighteenth century. For example, an Italian wine producer claims that "the Grand Duke of Tuscany drew the boundaries for the production zones [appellations] of important regional wines in 1716, creating the first official appellation in the history of winemaking."[16] With different qualifiers that support the legitimacy of their claims, regions of Hungary and Portugal also lay claim to the first appellation. Because wine is a traditional product, the age of winegrowing enterprises carries marketing clout, but whether and why wines from those regions were superior cannot be determined from the length of their tenure.

Some wine journalists attribute the origin of the terroir concept, and recognition of the dominant role of the soil, to the medieval Benedictine or Cistercian monks. The Cistercians produced wines of excellent reputation from vineyards in Burgundy and other areas, such as the Rheingau, including some famous for great wine today: Clos de Vougeot near Beaune, France, and Kloster Eberbach near Johannesburg, Germany, on the Rhine. The Cistercians, a spin-off of Benedictine monks, originated at Cîteaux Abbey as a reform movement seeking a more spiritual life through a stricter and more austere Benedictine form of Catholicism: simplicity, poverty, and living off of their own labor. According to wine journalists Hugh Johnson (*Vintage: The Story of Wine*) and Karen MacNeil (*The Wine Bible*), the Cistercians in Burgundy discovered terroir by paying close attention to which plots of land made the best wines. A lot of specifics about this discovery are provided, particularly in Johnson's *Vintage:*

> Their greatest contribution to wine was the concept of the "cru"; a homogeneous section of the vineyard whose wines year after year proved to have an identity of quality and flavor. They observed that differences of colour, body, vigour, and other qualities in the wine were remarkably constant from one patch to another. They made small batches of wine from separate plots, compared the scores of samples of tithe wines that came their way, and began to form a picture of the resources of the Côtes [hillsides]: which parts made a more aromatic wine, which more robust and rough, which suffered most from frost, which needed picking early; a whole data bank of information. Then they started drawing lines on the map; even building walls around the fields that regularly produced recognizable flavor.[17]

According to an essay in the *Journal of Wine Research*, the Benedictine and Cistercian monks observed "with remarkable astuteness the nuances derived from one plot as compared with another."[18] Wine experts report that the early monks were almost uncanny in their viticulture insight.

Outside of the popular wine press, the Cistercians have enjoyed an outstanding reputation of rapid expansion from their beginning in 1098, guided by a male-dominated order that employed an innovative approach (sending out a few monks to establish outposts, from which new monasteries developed), and made major contributions to agriculture and architecture with a dedication to new technology and quality. They were successful agronomists and became an important economic institution for a period that spanned approximately the period AD 1100–1400 in France, Germany, Britain, and parts beyond.

Historian Constance Berman, however, argues that the remarkable story of the Cistercian movement is a somewhat revisionist history, presented ex post facto by the Cistercians themselves. According to Berman, the surviving sources tell a different tale—one of a slightly later foundation, expansion by the absorption of loose monastic networks that already existed, and an order that had active and independent-minded female members.[19] In addition, "anonymous peasants needing more land for their families, hermits going out to live alone in the forest, and lords intent on getting a share of the profits of that expansion" were more important factors than the religious orders in agricultural developments, and most of these improvements had already been made by the twelfth century.[20] Secondary improvements promoted by the new monastic groups continued into the thirteenth century, and included the consolidation of fragmented old estates, the elimination of middlemen from claims to the produce of estates, the building and improvement of mills, and the early beginnings of selective animal breeding.

Although the terroir story has the Cistercians breaking down vineyards into small plots, the Cistercians actually developed an agricultural "grange" system, which consisted in no small part of consolidating small tracts of land into larger fields.[21] The research on this generally refers to arable land usually planted to grains and legumes, but the Cistercians consolidated vineyards as well. They enclosed multiple vineyards with a wall to create the largest vineyard in Burgundy, Clos de Vougeot.[22] The walling off of vineyards was not specific to Burgundian vineyards; there was a general movement from open- to closed-field agriculture during the High Middle Ages, and such closed vineyards were clearly established, and in fact were already being sold off in and around Paris, as early as the turn of the thirteenth century.[23] Berman and other medieval historians describe the beginning of the end of the Cistercian influence as arising from their movement toward farming as a business and as a means to collect rents, taking on conversi laborers, and relinquishing control over production practices.[24]

Geography scholar Tim Unwin acknowledges the religious order's long ownership and relatively advanced education as creating the opportunity for research. He even goes so far as to say the monks were "keen to experiment."[25] Perhaps the monks were advanced in experimentation. The monks may have had the necessary time, and the relatively large amount of labor needed to assist in such matters was available before the Black Plague (ca. 1350). The astuteness and detailed record keeping attributed to the monks would have been unusual, however. The documents that Berman found were the more necessary trade and tax records.[26] The motivation for investing labor in parsing smaller plots, and in keeping detailed records of each vat or barrel with tasting notes, seems lacking. At the time of the Cistercians, scientific or analytical thinking was not important. Medieval thinkers utilized a deductive method of thinking and often uncritically accepted the ideas of authorities, turning to the Bible or the ancients for answers. For questions dealing with the cosmos, one opened the works of Ptolemy; for medical matters, one turned to Galen. Other medieval thinkers practiced astrology; at that time experimentation was rare and not an accepted means of gaining true insight. The closest to an experimental science was early alchemy, a mixture of magic and chemistry.

The Cistercian-based story of terroir is told and retold in the popular press and on various wine-related websites. Yet that story is absent not only in Berman's writings, but also in the works of respected French historians Roger Dion, Jean-Robert Pitte, and Georges Duby, among others.[27] The apparent disparity between the popular legend and the academic research raises some question as to what actually took place and what was truly revealed at the time.

Whether or not the early Cistercians embarked on what is portrayed as a scientific study of winegrowing (hundreds of years before the Scientific Revolution) is not clear. Regardless, it's important to note that wines in the Middle Ages were usually quite bad by contemporary standards. There was no germ theory or Pasteur yet, and wine in the thirteenth century was necessarily drunk young. No attention was paid to vintage, and often what was served was of poor quality, even at rich tables. French poet and diplomat Peter of Blois (ca. 1135–1211) described in a letter the wine served at King Henry II's court: "The wine is turned sour or moldy—thick, greasy, stale, flat and smacking of pitch. I have sometimes seen even great lords served with wine so muddy that a man must close his eyes and clench his teeth, wry-mouthed and shuddering, and filtering the stuff rather than drinking."[28] In Paris, the price of wine was low when the supply of new wine was available each year, and increased as the supply of sound wine decreased because of spoilage.

Of course at the source, the Cistercian monks would have had the opportunity to drink unspoiled new wine, and may have distinguished differences that they attributed to vineyard soils. But even so, if a wine was of poor quality, was it because of the soil, picking unripe or over-ripe fruit, the yield, the variety mix (at that time, there were almost always mixed plantings), poor weather, the ox manure, or spoilage organisms? In addition to the wines being poor, if not altogether unrecognizable to today's tastes, it is unlikely that the Cistercians had the means to sort out the many potential causes of wine differences.

It's also worth noting that it would be a matter of special *dis*organization for grapes from a certain location to not be fermented together. The size of a plot of land that could be distinguished in the wine could be as small as that for a single fermentation. Hugh Johnson cites a report on a 1789 publication, in which an attribution to an earlier paper appears: "There was a priest called Claude Arnoux who wrote his *Situation de la Bourgogne* in 1728 and he said, 'those who opt to make the best wines only put in the vat the grapes from the single vineyard.'" Johnson appropriately asks, "Who was to say what a single vineyard was?" One key to the size of plots would be the size of the crushing, fermentation, and pressing vessels, because the characteristics of plots smaller than those that are fermented together would be obscured. Although there are some records of yields and transactions, units and vessel sizes at that time were conspicuous messes.[29] I could find no reference to what land area was usually fermented together during that time, let alone information on how the Cistercians deviated from popular practice because experience showed them it was prudent.

It is just as likely that the monks were motivated toward efficiencies, such as those created by vineyard consolidation, as well as larger presses and fermentation vats. Indeed, large presses remain as tourist attractions at some of the oldest and most prestigious of the monks' sites. Nevertheless, it is possible that in a few places, important vineyards with rolling hills, that the Cistercians made changes in the opposite direction from the well-documented consolidation—splitting up vineyards, which would allow for careful comparisons.

BOUNDARIES OF PRIVILEGE

It is informative to learn and understand the timing of the emergence of names that are associated with the best wines, where wine quality is attributed to terroir. Already mentioned were the medieval vineyards of Clos de Vougeot in France and Kloster Eberbach in Germany. By the early 1800s, Clos de Vougeot was already ballyhooed as producing the most prestigious Burgundy wine in the world, and for doing so for more than 300 years without introducing "new vines."[30] Although some Burgundy wines were held in the highest regard, they were less developed into trade, in part because of the limited river routes to markets. Indeed, among the most sought-after wines was Malmsey from Crete and Cyprus, a sweet wine that could hold up during shipping, perhaps having been boiled first.

The reputation for excellence of current preeminent vineyards and estates was established early, and this is especially clear in Bordeaux. With the English in control of southwestern France from 1152 to 1452, much winegrowing was directed at the English market. Authorities granted winegrowers the privilege of bringing duty-free wines into Bordeaux from a restricted area around the city. Wines produced outside this limit were subjected to taxes and could not be brought into the city before late fall or early winter. This provided the Bordeaux proprietors a monopoly, which was perpetuated until the French Revolution. "When, in the early 1900s, areas of appellation were designated by law, the winegrowers of the Department Gironde claimed again, and obtained the exclusive title to, the label 'Bordeaux,' while vineyards in nearby areas like Bergerac, which was outside the old area, were again forced to use different names for its wines."[31] Thus, from early on the boundaries of a wine region have been boundaries of privilege.

According to Clive Coates (*Grands Vins: The Finest Châteaux of Bordeaux and Their Wines*)[32] and Hugh Johnson (*The Story of Wine*), esteemed wine families in Bordeaux were already increasing their control of winegrowing in the sixteenth century, in part by the consolidation of landholdings. In Médoc, the desirable high land was consolidated first, as the rest of the remaining area was swampy lowland. During this time, the Pontac family was ascending in Bordeaux, with Arnaud de Pontac becoming mayor in 1505. In the 1570s, Pierre de Lestonnac began assembling land that would become Châteaux Margaux, while the Pontacs were acquiring other lands—one of whom, Arnaud de Mullet, became owner of what would later be known as Latour. When his son (who had assembled the lands that would become Latour in the 1650s) died, the Latour estate was inherited by the Lestonnac family, who were already owners of Margaux.

After the consolidation of the high land, the Dutch drained the swampy Médoc with contracts that began in 1599. In 1660, another Pontac served as senior judge of the region Guyenne (aka Gascony). By that time, the Pontac family also owned the château of Haut-Brion in the Graves region near Bordeaux. Although wine had been sold widely under the name of Pontac, wine was sold in England under the name of the estate (evidently for the first time) in 1663 when Londoner Samuel Pepys referred to Haut-Brion in his diary as having a good and most particular taste. In 1665, a Pontac opened a tavern with high prices in London. By giving the château name to the best wine and using family names for other properties, the Pontacs essentially had two brands—an important development in the marketing of wines. Hugh Johnson attributes the success of the Pontacs to a collection of good winemaking practices and marketing skills.

As early as 1725, the Bordeaux merchants classified their finest red wines as "First Growth," "Second Growth," and so on. The top handful of Bordeaux wine operations were essentially those that were established first and politically influential. Coates reported on the top-ranked wines from 1714–74 and included these same early producers.[33] Coates also noted that producers who were not top ranked, but would become so, were less well-connected politically. Many of the Second Growth operations had vineyards that were not yet planted when producers such as Haut-Brion and Latour had already established their high reputations.

Thomas Jefferson conducted wine tours during 1787–88, while serving as ambassador to France. His extensive notes were recently evaluated by John Hailman in *Thomas Jefferson on Wine*. The wine hierarchies of today were largely in place by the late eighteenth century, and Hailman shows that Jefferson recorded almost the same hierarchy of familiar châteaus in the Médoc.[34] Several wine rankings created by wine brokers in the 1800s show similar standings.[35] This structure was formalized in 1855 based solely on a price list, and is still in place today. Similarly, in 1861, the hierarchy of vineyards in Burgundy was first established, leading to the later designations of Grand Cru, Premier Cru, and so on. In Burgundy, it was vineyards, rather than estates, that were classified. There was some further formalization in the nineteenth century, although not yet to the legal standing that they have today.

The origins of terroir extend back, depending on your expert source, to the Egyptians 5,000 years ago, ancient Greece, Roman Italy, medieval Burgundy, or to eighteenth-century descriptions of top European vineyards. Based on several sources, the high reputation of specific vineyards or winegrowing houses in Europe was clearly in place by the late eighteenth century, in Burgundy, Bordeaux, and Rheingau, as well as in winegrowing regions in Italy, Hungary, and Portugal.

So how are we to understand or explain the different reputations for wine quality that are attributed to various places? In early times, perhaps the only available explanation for differences in wines (and wine quality) was the differences in the soils, which were considered the source of all plant growth and thus a logical explanation of the resultant fruit and wine flavors. There were few alternatives imaginable, except an attribution to the direct hand of God. Science was in its infancy, and no one was in a position to establish anything like an objective analysis of wines, based on reproducible observations of how the soil affects the plant and its fruit. Yeast and fermentation were not yet understood, so it would have been a big step to suggest that the winemaking techniques were key to wine quality. Thus, the staying power of the hierarchy and the mystical aspect of terroir in the face of myriad scientific advances are quite impressive.

One example of the soil-based story can be found in the notes of the philosopher John Locke, who visited France for five years and in 1677 wrote highly of Haut-Brion. He repeated the common story that wines differ dramatically between adjacent vineyards ("separated by only a ditch"), and commented that the quality of the wine depends first and foremost on the soil.[36] Tim Unwin takes this to indicate Locke's sound knowledge of wines and viticulture. Another way to interpret his writings could be that Locke, lacking sound and independent knowledge of wines and viticulture, accepted what the locals told him as fact. Indeed, Unwin argues elsewhere in the same report that Locke's position in society necessarily limited his access to fine wines, so one might reasonably assume a similar case for his knowledge thereof.

James Busby, known as the father of Australian wines, visited France 150 years after Locke. He heard the same refrain, but had a different take on it: "The limited extent of the first-rate vineyards is proverbial, and writers upon the subject have almost universally concluded that it is in vain to attempt accounting for the amazing differences which are frequently observed in the produce of vineyards similar in soil, and in every other respect, and separated from each other only by a fence or a footpath. My own observations have led me to believe, that there is more of quackery than of truth in this."[37] The traditional story has continued to thrive, although Busby's take could and perhaps should be given some more attention.

EARLY USE OF *TERROIR*: A PEJORATIVE IN WINEGROWING

The origin of the word *terroir* is evidently the Latin word for "soil"—*terra*—which was commonly used before the development of written French.[38] Although defined from its earliest written use as "soil" or "land," *terroir* was also employed in combination with wine flavor from very early on. The Maynard Amerine Collection at UC Davis claims the world's best collection of viticulture and enology material. A survey of this collection reveals an evolution of the use of *terroir* in written French and later in English, as well as in French and French-English dictionaries and wine books.

The Amerine Collection includes a copy of the first book on wine printed in French: *Deuis sur la vigne, vin et vendanges / d'Orl. de suaue; auquel la façon ancie(n)ne du pla(n)t, labour & garde est descouuerte & reduitte au present vsage* (On the vine, wine, and harvest: The old way of plant, labor, and storage as revealed by the present use), written by Jacques Gohory (1520–76) and printed in 1549. His use of *terroir* appears in conjunction with wine in a few sentences, and is readily translated as "land," as in "Italy boasts a Greek wine from the land (*terroir*) around Naples."

The same year, French lexicographer and publisher Robert Estienne issued his *Dictionaire francois latin*. He included the phrase *gouste du*

TERROIR. ſ. m. Terre conſiderée ſelon ſes quali-
tez. Les plantes, les arbres, ne viennent bien que
ſelon que le *terroir* leur eſt propre. Les ſaules, les
aulnes, les peupliers demandent un *terroir* humide &
mareſcageux; la vigne un *terroir* ſec, pierreux & de
roche; le bled un *terroir* gras & fertile. Le *terroir* des
landes ne ſe cultive point, parce qu'il eſt trop ingrat.
On dit que le vin a un gouſt de *terroir*, quand il a quelque
qualité deſagreable, qui luy vient par la nature du terroir où la vigne eſt plantée.

FIGURE 31. "TERROIR" ENTRY IN THE UNIVERSAL DICTIONARY, CONTAINING ALL THE
GENERAL FRENCH AND THE TERMS OF ALL THE SCIENCES AND THE ARTS (Furetière 1690).
"TERROIR. Singular masculine noun. Soil considered according to its qualities. The
plants, the trees, grow only on/in the terroir that is good for them. Willows, alders,
poplars need a humidified and swamp-like terroir; grapevine needs a dry, stony terroir;
wheat needs a fertile and fat terroir. . . . People say that wine has a taste of terroir, when
it has some disagreeable quality which comes to it by way of the nature of the terroir
where it is planted." (Trans. Sophie Mirrasou.)

terroir, "taste of the soil or land," referring in particular to a flavor of
wine, although it is not clear whether that taste was good or bad. In a
work published in 2000, *Le dictionnaire des façons de parler du XVIe
siècle: La lune avec les dents* (Dictionary of the ways of speaking in the
XVIth century: The moon with teeth), Pierre Enckell, journalist and
lexicographer of French expression, reports that *terroir* was often used
as a pejorative with respect to wine flavor, and this usage is borne out in
many references.

In the 1690 *Universal dictionary, containing all the general French
and the terms of all the sciences and the arts,* terroir is similar to soil,
and a taste of terroir in wine is considered to be a disagreeable quality
arising from that soil (fig. 31).[39] The *Dictionnaire de l'Academy fran-
caise, dedié au Roy* (Dictionary of the French Academy, dedicated to
the King, 1694) provides a similar definition. Thus, from our earliest
records, *terroir* is employed as a pejorative when describing wine flavor,
and that undesirable flavor is derived from soil.

In one 1650 French-English dictionary, the definition of *terroir* is
"soyle, manure, or land"; in this instance, synonymous with *manure*
itself. *Terroir* was defined in 1750 and 1843 dictionaries as "soil" or in
goût de terroir without indicating a positive or negative connotation.

Nevertheless, the pejorative use was common in dictionaries and wine texts in the seventeenth, eighteenth, and nineteenth centuries.[40] In 1866, wine writer André Jullien used *terroir* in this way to describe several wines: "The wines from other wineries have an unpleasant taste of terroir, and are devoid of body and spirit."[41] From his glossary entry: "*goût de terroir:* It communicates to the wine by the land (or soil) on which it was harvested" (in French: "Il est communiqué au vin par le terrein sur lequel il a été récolté").[42] For Jullien, this taste of terroir was not a good thing. Even the famed Chaptal referred to the *goût repoussant de terroir,* "the repulsive taste of terroir," and *goût de terroir très fort et très désagréable,* "a very strong and unpleasant taste of terroir."[43]

The negative connotation of the term *terroir* was still reported in the American press near the turn of the twentieth century. Consider the following, for example: "In Cognac, a terroir is the lowest ranking of a vineyard (premier, fins, bons, . . . terroir) which on account of its strong unpleasant taste cannot be employed except in small blending quantities."[44] In California, Frederic Bioletti translated the 1889 Italian booklet *The Wine* and reported it to California's emerging viticulture industry in 1892. The book is rich with descriptions of wines and their sources (e.g., "earthy taste—*goût d terroir;* a general term for flavors that 'are all in general disgusting or bad,' may come from wrong manuring or other plants/weeds. Generally, the earthy taste is not found in high-class or fine wines.").[45] *Terroir*'s use in reference to a wine's failings continued at least up to a 1928 travel report in southern France: "The local wine 'lacks breed' and even with age doesn't lose its *goût de terroir.*" The wine "pays in quality for the richness of the soil."[46]

It is clear that *terroir* and *goût de terroir* were often employed to refer to an off-putting flavor that was considered to be local and soil derived—often in a quite literal sense. We will return to the history of the use of *terroir,* but first let's catch up a bit on how the grapevine was once thought to interact with its environment to make a grape.

CONNECTING TERROIR TO THE GRAPEVINE

At the dawn of civilization, humans learned that manure or other organic matter placed on or in the soil led to increased plant growth, but this observation did not become a theory of plant nutrition until after the advent of scientific inquiry, and even then it was a "tough row to hoe." At one point, plants were described as having "spongioles," sponge-like endings on the roots that soaked up nutrition, but alas, like the black bile of humorism, these "spongioles" were never observed and faded from existence, even as a concept.[47]

Flawed Concepts of Plant Nutrition

The medieval sense of plant nutrition encompassed three problematic concepts that impeded progress in understanding the nature of grapevines. These concepts live on in the "wines as an expression of the soil" belief that is inherent in the popular terroir explanation.

Most important among these concepts is transmutation—a precursor to chemistry. The alchemist tried to *transmutate* various substances into gold, but transmutation was also employed as an explanation for the creation of plants from soil or water (and thus for leaves and berries). Even further, it was once universally believed that transmutation could lead to the metamorphosis of one plant species into another. Consistent with the terroir concept, berries were thought to come from transmutated soil, and therefore directly reflected the soil in which the roots grew.

Vitalism, the second flawed concept, is the notion that something nonmaterial, a soul or life force, must be added to inorganic matter in order to produce life, and that life is (at least in part) the product of this mysterious and unknown substance or energy. In other words, vitalism is the idea that life is due to a force beyond the ordinary workings of chemistry and physics. For many, the vital force was contained in the humus (decaying organic matter in the soil), and this uptake of humus was what allowed one generation of living plants to succeed from another. Vitalism per se can be traced back to Galen, but the concept really arose as a reaction to the mechanistic description of biology that emerged during the Enlightenment. Vitalists believed that the laws of physics and chemistry alone could not explain life functions and processes. A recent history of vitalism in botany declared: "Indeed, for many contemporary philosophers, consciousness, if not cognition, still holds out against materialist reduction. But not life. Vitalism is dead and buried as a serious theory of the 'vital' phenomena: self-motion, nutrition, growth and reproduction."[48] Vitalism has been declared dead many times in academia, but persists in some lay circles.

The final flawed concept, teleology, is an extension of vitalism that claims phenomena move toward certain goals of self-realization, a logic that in effect puts the consequence before the cause. From Aristotle through the Middle Ages, the explanation for apples falling from the tree was that apples fall down and not up because it is their nature to fall down; the bird has wings in order to fly (yet one must have a need to fly first, then one can have wings). Telelological thinking is expressed in anthropomorphic phrases that impart a will to plants; the Chardonnay grapevine is *trying* to make good Chardonnay wine, and can do so when it is in its proper soil.

According to eminent biologist Ernst Mayr, all three concepts (transmutation, vitalism, and teleology) had to be rejected in order for biology to emerge as a functioning science.[49] Fundamental to the new "scientific" method was a revolutionary change in the approach to understanding nature. Aspects of teleology remain in debate today in biology, but the original notion that we could understand and explain design and behavior in nature based upon what we view as the ultimate application has had to take a back seat to experimental biology.

In the new scientific approach, measurements were recognized as more compelling than intuitive constructs, the word of the church, or the writings of the ancients. For example, Galileo stopped trying to explain *why* objects moved and instead started measuring *how* they moved. About 100 years after Galileo, Stephen Hales, known as the father of plant physiology, gave the following sage advice: "The likeliest method to enable us to make the most judicious observations, and to put us upon the most probable means of improving any art, is to get the best insight we can into the nature and properties of those things which we are desirous to cultivate and improve."[50] Hales was an important convert to measuring things, rather than reporting his impressions or intuitive explanations. He worked for several years to find a circulation system in plants that was parallel to that found in animals, but his measurements allowed no such conclusion. Hales made various scientific contributions, and his finding that the analogy approach to studying plants and animals had limits aided in the movement away from Aristotelian understanding and method in plant biology.

Terroir in the Early History of Plant-Soil Interaction

One of the earliest cited advances in thinking about plant nutrition is that of French engineer and noted potter Bernard Palissy (1510–89), who reported that "manure is carried to the field for the purpose of restoring to the latter a part of what had been removed."[51] Not a great leap from today's vantage point, but a very important step toward resolving just what plants need and obtain from soils. Note that even this insight creates a problem for the terroir explanation, because it implies that the part that matters to the grapevine is something the farmer brings (or can replace).

Palissy's idea that plants deplete soil nutrients (and that these nutrients can be replaced by fertilizers) was later confirmed but was not incorporated into the terroir worldview of winegrowing. About a hundred years after Pallisy's insight, another problem for terroir arrived with the discovery that plants could essentially grow without soil. In a famous experiment, Jan Baptista van Helmont (1579–1644) planted a willow tree in potted soil that he had weighed, and then covered (to keep it from losing or gaining soil), and cared for the willow for a period of five years, watering it with rain or distilled water.[52] After five years, the willow had gained 169 pounds, but the soil lost only 2 ounces (fig. 32). He attributed the plant growth to transmutation of water, because his experiment had proven that it was not due to transmutated soil.[53] The wine industry could have set its claims on the local water, much as Coors beer did, but stuck with the soil and terroir instead.

Van Helmont also wrote that when 62 pounds of oak charcoal were burned, they would yield 61 pounds of gas and 1 pound of ash.[54] Thus, van Helmont knew that dry plant matter released large amounts of carbon dioxide upon burning. Had he not been so dogmatic about water transmutation, he might have used his data to conclude that fresh plant matter consisted largely of water, but that dry plant matter consisted mainly of carbon dioxide gas and a small amount of soil minerals. That kind of conclusion would have advanced plant biology by a century.[55] Perhaps the obvious fact that plants were not excavating huge holes beneath themselves contributed to the popularity of van Helmont's interpretation. Water transmutation became a prevailing wind similar to terroir, an idea to which many scientists clung to well into the eighteenth century, despite evidence to the contrary.

In 1699, John Woodward grew spearmint in rainwater, river water, and conduit water. He found that plants grew best in water that contained the largest amounts of suspended material. Woodward concluded that plants are made up of constituents in the water derived from soil and not water alone. He was right (at least in that it was not water alone and that the other stuff is important), but how the plant grew remained a mystery.

After van Helmont and Pallisy, perhaps the earliest move away from the soil (or water) transmutation explanation for plant growth was made by the French scientist Edme Mariotte (1620–84). While more famous for the discovery of the blind spot in the eye, Mariotte also

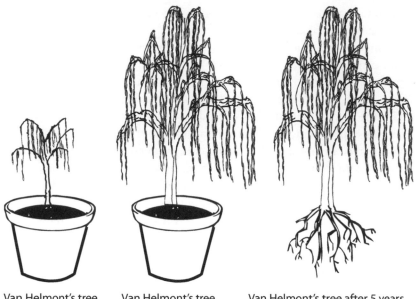

| | Van Helmont's tree when he planted it | Van Helmont's tree after 5 years | Van Helmont's tree after 5 years out of its pot to show the roots |

	Mass at start (kg)	Mass after 5 years (kg)	Change in mass (kg)
Tree	2.27	76.74	74.4
Dry soil	90.72	90.66	-0.06

FIGURE 32. VAN HELMONT'S WILLOW TREE AND SOIL, PICTURED BEFORE AND AFTER FIVE YEARS OF GROWTH IN A POT. The van Helmont experiment showed that a tree could grow more than 70 kg with very little soil being "consumed" or transmutated.

suggested that a plant is comprised of certain constituents, such as salts, water, and so on, that are in turn comprised of more simple elements that have united together. He was onto something important. Mariotte remarked that these soil-derived elements had no "consciousness by which they seek to unite together," but they had some natural properties that led to such interactions.[56] (This uniting would be the physiological processes that give rise to growth and ripening.) He was describing naturally occurring mineral nutrients, as opposed to the preformed miniatures of Aristotle or corpuscles of organic matter.

In a description of interactions within the seventeenth-century Parisian Royal Academy of Sciences, historian Alice Stroup found that Mariotte "argued instead that plants developed their parts and properties gradually, as the result of an interaction between the plant and its sap."[57] This sounds more like a metabolic synthetic process familiar to today's plant biologists. According to Julius Sachs's *History of Botany*,

> With this first hypothesis Mariotte places himself in opposition to the Aristotelian doctrine with its entelechies [vital forces] and final causes [teleological explanations] which prevailed at that time among botanists and physiologists upon the firm ground of modern science with its atoms and its assumption of necessarily active forces of attraction. . . .
>
> It is interesting then to see how vigorously Mariotte exposes the incorrectness and absurdity of this idea (preformed corpuscles) though he has no new discovery to help him. In his third hypothesis he maintains that the salts, earths, oils and other things which different species of plants yield by distillation are always the same and that the differences are due entirely to the *way* in which these coarse principles and their simplest parts are united together or separated and he proves it thus: "If a *bon chretien* (Williams variety) pear is grafted on a wild one the same sap which in the wild plant produces indifferent pears produces good and well flavored pears on the graft, and if this graft has a scion from the wild pear again grafted on it the latter will bear indifferent fruit."[58]

Mariotte expressed that the nature of what plants obtain from the soil, and what the plant then does with this nutrition, is far more complicated than the concept that little "bits that want to be Chardonnay" are taken up from the soil, and under the right conditions are allowed to become the best Chardonnay they can be. Of course, there was a prevailing wind against this new thinking. Nehemiah Grew was an important plant anatomist in the early 1700s who still held to the uninvestigated idea that different plants require different "corpuscles" (the preformed substances of plants) for their formation and nourishment.

Following Mariotte, scholars divided into an interesting pair of camps—those favoring the soil mineral particles and those favoring the organic humus as the source of plant nutrients. Jethro Tull (1674–1740), the agricultural thinker credited with inventing the seed drill, was of the mineral camp. He believed soil mineral particles directly entered the root, and that tillage and organic matter both contributed to making finer mineral particles, which entered more readily and, therefore, produced better growth.[59] Those who believed plants took up minerals downplayed the role of organic matter to something that aided in a kind of soil digestion, helping to make smaller and smaller mineral particles.

On the other side of the debate, organic matter of the soil, or the soil humus, was regarded as the chief nutrient for plants and the major source of soil fertility. In the Humus Theory, roots were believed to extract the humus from the soil and to transform it into plant substance by combining it with water.[60] The proponents of the Humus Theory generally did not take into account mineral substances such as potassium, phosphorus, magnesium, and so on. Some of the proponents of the Humus Theory believed that the soil minerals were not essential for growth; they were believed to act as stimulants, or existed to enhance the decomposition of humus, rather than as nutrients. Others looked upon minerals as mere accidental plant constituents, or as the skeleton substances of plants somewhat similar to the bones of animals. According to this theory, plants feed upon substances that are similar to them in nature, and plant nutrition was still considered similar to animal nutrition in that both plants and animals feed on complex organic bodies.

Despite the problem created by growing plants essentially in water, and the benefits for crops realized by the invention of chemical (inorganic) fertilizer (referred to as "philosopher dung" and brought on by manure shortages during the Thirty Years' War, 1618–48),[61] the belief that humus was the food of plants was upheld well into the nineteenth century and supported by renowned chemists, including Sir Humphry Davy (1778–1829), who claimed that experiments on himself indicated that inhaling nitrous oxide was an effective cure for wine hangovers.

Yet plants are different from animals in important ways, including their source of energy and mass. The scientific foundations of the Humus Theory were proven false by Carl Sprengel, who showed (in 1826 and 1828) that plants live on salts and mineral nutrients that are dissolved in water, and by Justus von Liebig, who published similar findings, as well as a classic 1840 book outlining the concept that became known as "Liebig's Law of the Minimum."[62] For many students of agricultural history, Liebig's work launched modern agriculture. Liebig explained the Law of the Minimum by describing the various elements required for mineral nutrition as individual staves in a barrel, where the shortest stave (the limiting nutrient) determines the useful volume of the barrel (the capacity of the soil to support plant productivity).

Liebig became the dominant figure in plant nutrition and discovered nitrogen as an essential plant nutrient, but he incorrectly believed carbon was absorbed by roots, nitrogen was absorbed from the air, and that a vital force regulated the plant cell metabolism. Sprengel and Liebig showed that nutrients are taken up from the soil solution, not as soil mineral or humus particles (fig. 33); however, the specific essential nutrients (those that plants absolutely must to have in order to grow, reproduce, and make fruit) were not worked out until the early twentieth century.

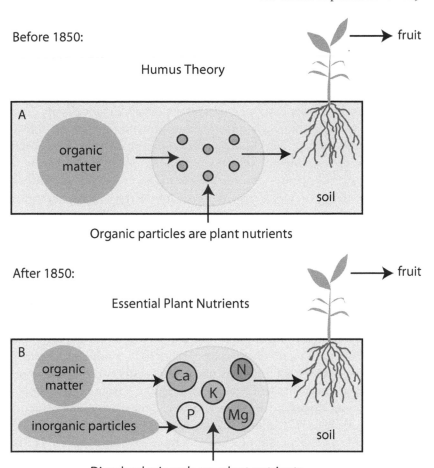

Before 1850:

Humus Theory

A

organic matter

soil

Organic particles are plant nutrients

fruit

After 1850:

Essential Plant Nutrients

B

organic matter

inorganic particles

Ca   N
   K
P   Mg

soil

Dissolved minerals are plant nutrients

fruit

FIGURE 33. THE TRANSITION FROM HUMUS THEORY TO CONTEMPORARY PLANT MINERAL NUTRITION. (A) In the Humus Theory, plants consume organic matter (humus) as food particles. (B) Today, we know that plants absorb dissolved mineral nutrients from the soil solution. The mineral nutrients are derived from organic and inorganic sources in the soil, and the plant does not distinguish among the sources of those nutrients.

An Important Role for the Aerial Environment in Making Winegrapes

Once it is understood that berries are not made by a plant "eating" soil, the question "What makes a berry?" becomes more complicated—and interesting. A berry is comprised of about 5 percent "ash"—a catchall term for the incombustible mineral content derived from the soil. For plants to utilize essential soil nutrients efficiently, the additional factors of light, heat, and water must be adequately supplied. Some of those inputs are used in nutrient uptake, but most are involved in creating the rest of the plant (including the berry) using the products of the process of photosynthesis in leaves. Photosynthesis extracts carbon dioxide from the atmosphere (and releases oxygen) to create the building blocks of sugars, and from those amino acids and other compounds comes the stuff of berry growth and ripening. But recognizing this process was a long time coming. Andrea Cesalpino was a contemporary of van Helmont and an important early botanist, who, in *De Plantis Libri* (1583), claimed the function of the leaves was to provide shade for the rest of the plant (others thought the leaves might excrete waste). It was not until the 1700s that it began to be suspected that leaves might have a role in plant nutrition.

In *Vegetable Staticks* (1727), Stephen Hales demonstrated that plants absorb air, and suggested that plants derive nourishment from the atmosphere through leaves: "May not light also, by freely entering surfaces of leaves and flowers contribute much to ennobling principles of vegetation." The roles of light, air, carbon dioxide, and water in photosynthesis were slowly resolved by the work of many European scientists. In 1780, the English chemist Joseph Priestley conducted a series of experiments that revealed that plants could "restore to a considerable degree of purity air that had been injured by the burning of candles."[63] The prevailing theory at the time was that flames are extinguished in airtight space because the air saturates with "phlogiston" (an imagined kind of bad substance). So, too, when a mouse was sealed in a jar, the air saturated with "phlogiston" and the mouse would weaken or perish.

Priestley placed a mint plant into an upturned glass jar of phlogisticated air for several days. He found that with the plant present "the air would neither extinguish a candle, nor was it all inconvenient to a mouse which I put into it."[64] Benjamin Franklin visited, and from experiments they conducted together, he noted, "We knew before that putrid animal substances, when added to the earth, produced sweet vegetables; and now it seems putrid air when mixed with plants similarly produce sweet vegetables."[65] By analogy to fertilizing, Franklin concluded that the act of the putrid air was to fertilize the plants, and that by deduction, the restoration of the air by the mint plant arose from removing something from the air. Priestley thought the same. He almost discovered that plants consume carbon dioxide and produce oxygen, but he could not recognize oxygen as such because he still believed in "phlogiston."[66]

The French chemist Antoine Lavoisier competed with Priestley and other contemporaries. Lavoisier argued correctly against the water transmutation idea, but by his own admission he was avid for glory. He evidently tried to repeat others' experiments and take credit for their discoveries, including some of Priestley's work. Nevertheless, Lavoisier terminated the phlogiston theory when he discovered oxygen and oxidation, and he is given credit for putting the four elements (humorism) to bed by showing that things in nature should be considered to be made up of those elements that they could be broken down into. (Lavoisier was executed during the French Revolution for being a monarchist sympathizer. According to legend, the judge who pronounced his sentence said, "The Republic has no need for scientists.")

It fell to a Dutchman, Jan Ingenhousz, who served as court physician to the Austrian empress, to make the next major contribution to the mechanism of photosynthesis. He had heard of Priestley's findings, and a few years later spent a summer near London conducting more than 500 experiments, in which he discovered the major role that light plays in plant function: "I observed that plants not only have the faculty to correct bad air in six to ten days, by growing in it . . . but that they perform this important office in a complete manner in a few hours; that this wonderful operation is by no means owing to the vegetation of the plant, but to the influence of light of the sun upon the plant."[67] Ingenhousz's discovery was in 1778, and by 1782 it was shown that carbon dioxide (CO_2) is taken up from the air during photosynthesis. The final contribution to the story was that of a German surgeon, Julius Robert Mayer (1845), who recognized that plants convert solar energy into the chemical bond energy found in the sugars that are products of photosynthesis.

With that, the leading botanists had a good start on how the soil and the aerial environments impact plant growth, but there were still few insights into the grapevine itself and how it made a good berry, let alone a good wine.

Recognizing the Variety's Role in Growing Grapes

It is standard crop production practice to seek out, develop, and plant varieties suited to the environment. Agronomists call this growing an adapted variety. Species adapt to particular environments in the wild by default—nonadapted individuals and their nonadapted genes are selected against, leaving the adapted individuals to reproduce. In agriculture, humans select individuals (dogs, sheep, grapevines) for traits that we desire. Modern breeding, in which varieties are crossed and desirable progeny retained, is a twentieth-century development. Gregor Mendel published his work on inheritance of traits in 1866, but it was not until 1900, sixteen years after Mendel's death, that scientists in Holland, Germany, and Austria rediscovered his work independently. Although plant hybridizations were made earlier, Mendel's work led to the field of genetics, upon which modern breeding is based.

Several important wine varieties can be traced back to the sixteenth or seventeenth century, so it is unlikely that those varieties arose from a conscious effort to combine good qualities of different parents to better meet the requirements of a given environment. In 2000, Professors Carole Meredith (UC Davis) and Jean-Michel Boursiquot (École Nationale Supérieure Agronomique in Montpellier, France) conducted genetic tests to identify the origin of the popular variety Syrah. They debunked myths that it originated in the city of Shiraz (once Persia, now Iran), or that it came from Syria or from Egypt via Syracuse. They determined Syrah was the offspring of parent varieties that probably came together in France well before people learned how to intentionally make genetic crosses. Pinot noir has an unclear history thought to reach back nearly 2,000 years.[68] Meredith also demonstrated that Pinot noir is one of the parents of Chardonnay, but it remains unclear just how Pinot noir emerged as a distinct variety.

Regardless of the origin of a variety, we depend on the genotype of each variety to deliver certain characteristics or traits that we associate with that variety. That's why we grow them. Agricultural selection may not make a genotype better adapted to survive in the wild, but we call agricultural varieties "adapted" when they produce the way we want, in the place we want. The role of genetics and variety was largely unknown and inaccessible when major European wine producers rose to prominence and established their reputations based on terroir. The role of the grapevine variety is well known today, yet the popular version of terroir and fine winegrowing remains focused on the place, usually emphasizing the soil.

Klebs' Concept: Nature and Nurture

The rediscovery of Mendel's work helped complete a series of important developments in botany in the late nineteenth century. The nature of plant nutrition from the soil was not as food, but as dissolved solutes in the soil water; and it is the process of photosynthesis taking carbon dioxide from the air that is the primary source of plant (and berry) mass. Cell theory, the idea that all plants are composed of cells, was introduced in 1838 and remains a foundation of biology today. The membrane theory of cell physiology, which proposed the fundamental role of cell membranes in controlling traffic into and out of cells, was proposed by the German botanist Wilhelm Pfeffer in 1877.

These advances allowed a synthesis of concepts: all observable traits (the phenotype) of an organism are the result of internal metabolism, which is, in turn, controlled by heredity (the genotype) as influenced by the environment. This basic truth of biology is sometimes referred to as Klebs' Concept (discussed in the introduction and illustrated in fig. 3). Once a variety is planted, the genotype is essentially fixed, and the whole of grapevine growth and development (and therefore winegrape quality) results from the variety interacting with its environment, as expressed in the physiology of the plant. Although it is often taken for granted today, this fundamental truth was known only by leading scientists of the day, when terroir-based appellations were being established in the early twentieth century.

Recall Edme Mariotte's comment on what makes a good eating pear: "If a *bon chretien* (Williams variety) pear is grafted on a wild one the same sap which in the wild plant produces indifferent pears produces good and well-flavored pears on the graft, and if this graft has a scion from the wild pear again grafted on it the latter will bear indifferent fruit." The same rising sap in the stem, regardless of the soil, gives rise to fruit qualities in each grafted shoot depending on the genotype of the grafted material. This was important and insightful in about 1680, well before Darwin or Mendel, but should not be news today. In fact, in a creative set of experiments in which just the clusters from one variety were grafted onto shoots of other varieties, Simon Robinson's group in Adelaide, Australia, showed that monoterpenes (important flavors of Muscats) are made in the berries themselves and are not delivered from the leaves outside the clusters. At UC Davis, we followed their lead and showed the same results for the synthesis of the green bell pepper/veggy–smelling MIBP.[69] Thus, what we know today of berry development indicates that berry colors and flavors (anthocyanins, tannins, norisoprenoids, monoterpenes, etc.) are synthesized in the berry and not transported to it from the soil or the leaves.

Some traits are more environmentally sensitive than others, subject to modification by the variety's interaction with weather, soil, and management practices. One often hears that Chardonnay is good for expressing terroir, and it may be that Chardonnay fruit composition is more environmentally sensitive than some other varieties, but this has received little experimental attention. How do these experts distinguish between variety-derived and soil-derived flavors? The exclusion of the variety from the explanation of a wine's flavor is a radical position. Moreover, and much more problematic logically, is that in order to know that a wine expresses terroir, a prior knowledge explaining the taste of terroir is required—who figured out what the taste of terroir is, and how did they do it? Let's not forget that the taste of terroir was considered to be an undesirable thing for a long time.

In focusing on the soil, or even the expanded version of terroir that includes more of the grapevine environment, the traditional Old World view diminishes the role of the winegrape variety (as well as everything that happens in the winery). Much of the international disagreement over wine labeling arises from the EU preference for naming an appellation (placing an importance on place to the exclusion of the name of the grape variety) and the United States (and other non-EU wine producers such as Australia) holding to the importance of explicitly including the grape variety on the label. The traditional view is that it is the place, not the variety, which determines the wine; however, this view ignores the biological constraint that the variety puts on the grape and wine. To be fair, imbedded in that view of place is another tenet of the terroir and appellation concept: the idea that the long history of Old World viticulture has imbued estate owners or appellation members with a special knowledge of which variety should be grown where, and that varieties in use are already known to be well suited to a place. In this regard, the diminishing of variety could be construed more as taking the variety as a given.

It is curious that the Old World terroir explanation of winegrowing leans on the soil, whereas the pioneering work describing the California winegrowing regions by Amerine and Winkler at UC Davis emphasized the aerial environment (i.e., climate, especially heat accumulation). Both traditions emphasize the adaptation of the variety to the environment; however, in some cases, the variety becomes more important than either soil or aerial environment, as with Barbera's high acid or Petite Verdot's high color. There is currently much more (and increasingly clear) empirical evidence of the roles of the aerial environment, light and temperature, in fruit ripening. This may be because the aerial environment has more impact, but it could also be in part because it is much easier to experiment by changing light and temperature than soils.

The preceding account of the early development of crop science and soil-plant interactions followed the leading edge of understanding of how plants make fruit up to about the turn of the twentieth century, which is when terroir-based appellations began to appear. What wine-growers and consumers could have understood about plants, soil, and berry development necessarily lagged behind when the reputations at the top of the French wine hierarchy were being established and attributed to their soils.

FLAVORS FROM THE SOIL

Of course, it *could be* that the taste of the wine, both good and bad, is indeed primarily a result of the soil. Although there are believers in the academic world, that popular view is not supported by increasing knowledge of how the soil makes or contributes wine flavors. The soil story was necessarily prominent in Burgundy in 1800s, but it was also known at that time that the best German wines were grown in the Rheingau on highly varied soils.[70] Chaptal was part of Lavoisier's anti-phlogiston cadre, and wrote extensively on vine biology. He also argued against soil as the primary factor of wine quality. His logic was sound: the soils that made fine wines were highly varied. Indeed, this is an important and overlooked fact that is even more clearly true today, with more extensive vineyard plantings and improved knowledge and surveys of the soils they grow on.

Descriptive soil studies have evaluated several agriculturally relevant soil characteristics associated with various "classed" wines in France. Several studies by Gerard Seguin of University of Bordeaux agreed that Grand Cru soils in Bordeaux are richer in organic matter, potassium, and phosphorus.[71] One might hypothesize that one or more of these factors contribute to the Grand Cru being Grand Cru. The soil differences, however, were evidently due to the contribution of humans, and not an inherent aspect of the place:

> These soils are richer because the Grand Cru vineyards owners have always brought organic matter and the other elements (manure, compost) in the soil even if the price of the wine went down.
>
> It is not because these soils are richer in organic matter, potassium, and phosphorus that they are Grand Cru soils, but it is because they are Grand Cru soils that they have been enriched in those elements by human contribution.

These differences in chemical properties between the Grand Cru soils and the others are not intrinsic because those differences probably didn't exist in 1855 during the classification.[72]

Thus, the soils of different classed growths are interpreted as being different because of farming practices—which can be conducted in any vineyard, regardless of location.

Survey studies of this kind are appropriate, but one must be careful in interpreting them, because they involve no treatments or testing of hypotheses, and a circular reasoning is sometimes involved: the best wines are made here, the vines are planted on these soils, and therefore these soils are the best for these wines.[73] It is difficult to experiment with soil type, but how were other soils tried or tested before conclusions were drawn on the best soils? Where are the records of the explorers who searched the world over, testing various soils for Pinot noir, Cabernet Sauvignon, Chardonnay, Riesling, Tocai Friulano, or Tempranillo, only to find the best soil was back in home in France, Germany, Italy, and Spain? A more likely history is that the varieties developed were adapted to their conditions and cultural predispositions. This leaves open the possibility that the varieties could be well (or even better) adapted to other environments.

Rocks and Grapes

One could get the sense from the popular press that both geological and soil-derived flavors are well-established facts. Today's popular wine books are typically a survey of wine regions, rich with attribution to terroir's influence on wine. Vineyard slope and climate are acknowledged, but quickly followed by descriptions of regional geology, with an attribution of wine style to the soil and the bedrock beneath the soil.

The rock explanation is promulgated in a series of wine books that contain more detailed descriptions of the geology than are found in most popular wine writing. *Terroir,* written by geologist James Wilson, and its French predecessor, *The Wines and Winelands of France,*[74] survey France's winegrowing regions. Wilson was searching for something that would validate the geology-based wine concept, and he toyed with the idea that iron levels in the soil were important. He may have picked up this idea while touring in France, but the connection between iron in the soil and flavors isn't there—at least not yet.[75] A California version of the geology tour, *The Winemaker's Dance,* appeared in 2004. Each of these surveys displays academic rigor with respect to the geology, but each implies that wine flavor comes from the rocks, and fails to make a meaningful connection between the rocks and wine flavor, flaws I attribute to an avoidance of biology and the fundamental truth of Klebs' Concept in particular. Each work bypasses the soil to some extent, and the grapevine for the most part, as well as the winemaking.

Historically, much has been made of chalky soils (when present) in France, Germany, and elsewhere. The eminent viticulturist Jules Guyot wrote in 1861, "The limestone is chalky soil that gives the juice the most outspoken and most free taste of terroir."[76] The chalk is said to give a special flavor to Franken wines,[77] and often a chalky or mineral taste is attributed to wines of vines grown on these chalky soils. This calls for an a priori knowledge of the taste of terroir or the taste that a chalky soil imparts. The chalk of Champagne is not present in Napa or Sonoma Valley, but the sparkling wines are often considered indistinguishable.[78] It is generally true that grapevines do well on calcareous soils, but it is probably more clear empirically that chalk deposits are good for holding oil reserves, than for flavors imparted to Chardonnay or other wines.

At the same time, there have always been observers whose own experience led them to question the conventional wisdom. British wine writer Cyrus Redding wrote in 1851:

> There is an obvious difference in the produce of vines grown upon particular soils, but they do not alter sensibly the character of the plant. The vines grown upon calcareous or chalky soils are not exclusively designated, any more than such as flourish upon those which are volcanic, and therefore they cannot be thus classified. The best dry wines seem to be intimately connected by nature with a soil more or less calcareous; the sweet are not thus remarked, but provided there be sun enough to mature or shrivel the grape, are produced on every kind of soil.[79]

Perhaps it is prudent to point out here that a soil is a "three dimensional body comprised of solids, liquids, and gases on the surface of the Earth that serves as a natural medium for the growth of land plants."[80] Thus, soil is the stuff that interacts with and supports plant growth.

Geology is the science that deals with Earth's physical structure, substance, history, and the processes that act upon it.[81] What can geologists contribute to the understanding of the berry and wine composition? Certain geologists could identify what kinds of rocks soil particles were derived from (as well as their elemental and mineral composition), but the role those particles play in fruit and their wines is a bit distant. The geologist is more likely to use the plants that are present to infer something about the soil profile and its mineral origin than to use the soil profile to infer something about plant growth, let alone fruit composition.[82]

The study of soils and their interactions with plants was the essence of academic agronomy for most of the 20th century; modern farmers know their soils. Studying soils, distinct from geology, has given rise to whole departments in academia, PhDs in soil science, and scientific societies around the world. Soil scientists have developed assays for nutrients and water in soils that inform about plant growth and productivity. The agricultural soil scientist and the plant ecophysiologist have in their crosshairs the physical and chemical properties of soil (including the organic components) that affect fertility, water availability, and plant growth. Their interest is not in where the soil mineral came from, but whether it is available to the grapevine and how it will contribute to its growth and fruit composition.

Let's also clear up one idea about soil genesis that is implicit in talking about rocks and wines—soils are not simply the broken-up bits of the rocks beneath them. Although many soils are derived largely from the weathering of the rocks beneath them, soils are determined by five factors: climate, organisms, parental material (the rocks below), topography, and time. Climate is usually considered to be the primary factor in the development of soil profiles. Soils can have inputs from outside ecosystems—wind brings dust and other deposition, floods and erosion move particles into or out of an area, and so on. Alluvial soils (often good for agriculture) are delivered to the site from elsewhere by water, and thus are often not derived from the rocks underneath them. Other soils arise from glacial till, which can be mostly comprised of material transported from elsewhere as well. Furthermore, none of the soil organic matter or nitrogen that the plant takes up comes from the rocks, and agricultural soils usually have outside inputs for the purpose of growing the crop.

The minerals derived from rocks may represent a relatively small part of the soil's impact on plants, except where the mineralogy makes farming difficult (fixing nutrients in an unavailable form, containing excessive levels of boron, sodium, or aluminum, or being deficient in difficult-to-correct micronutrients). On the other hand, the soil's physical attributes, particle size distribution, and the organization of soil particles—regardless of the specific mineral basis—are almost always important to root growth and water supply.

Jake Hancock, former professor of geology at Imperial College London, wrote the section on geology in *The Oxford Companion to Wine*.

Although he found the study of soils inaccessible and made many connections between geology and wine (that I find speculative),[83] he had not lost his thinking cap while developing an appreciation for wines. "He raged particularly against the concept of terroir (the idea that geological and climatic characteristics of a district impart unique flavors to a product grown there) which he saw as mostly superstition and pseudoscience."[84] A host of research symposia have been held that include geology reports on the wine regions of the world and the geological basis of wine flavors. There is hardly a geologist to be found at an apple or tomato meeting. Is the grapevine unique, or is it the refreshments? The simple fact is that grapevines have next to no interaction with rocks.

Tasting the Soil?

Another strange approach is that of tasting the soil. Hugh Johnson wrote that "the Burgundian way, and also the German way, introduced by the monks, was based on terroir, meaning the soil—already in some places so finely discriminated that, as Lalou Bize-Leroy once memorably said, she believes the monks actually tasted it. And they would need to, because the Côte d'Or does not give its secrets away."[85] There is some reference to soil tasting by the Romans, and perhaps the monks of the Middle Ages did this too. Tasting the soil probably falls into the category of pseudoscience or even quackery. Wine writers are often poetic, and it is not clear how serious Johnson was when he said they would need to taste the soil, but these comments do imply a limited perspective on how one might go about gaining understanding of the role of soil in grapevine growth and fruit development.

I have a report from a local grower newsletter in Oregon that describes how they imported two pros from Burgundy to provide advice on their Pinot noir. According to the newsletter, the two consultants actually tasted the soil (and gave their approval). But what does Ms. Bize-Leroy or the hired pros from Burgundy think could possibly be obtained from this—a flavor that would appear in the grapes? One could potentially discern high salinity and low or high pH from tasting soil, but that is not much information on the basis of which to discern the potential for fine winegrowing or on which to build a management strategy.

And yet soil tasting is promoted and apparently on the rise in the pop world: "It used to be," writes William Bryant Logan in *Dirt,* "that a good farmer could tell a lot about his soil by rolling a lump of it around in his mouth."[86] Logan searched for soil tasters, but found them hard to come by. One such taster was Bill Wolf, an organic pioneer who worked under Buckminster Fuller and who used to eat soil, until his doctor made him stop. Logan writes as though we have lost our way by using modern analytical tools, claiming that "science cannot really model such a complex dynamic (as a soil)."[87] A string of soil-tasting events have been held recently in California, South Africa, and the United Kingdom, often with the participating soils displayed in wine glasses, just in case you are slow on the implication. For example, "Taste of Place" was an artist's traveling construct/participation event that involved smelling soils and then tasting the "continuity" with products from those soils.[88]

These soil tasting events are not representative of the wine industry, but are in line with the fanciful notions about the soil that live on in the world of wine. If the flavors came directly from the soil, the leaf should have flavors that are similar to those of the berry, but more concentrated, because so much more soil solution is transported to the leaves than to the berries. But it doesn't work that way; leaves taste like leaves. Rather than soil-human interactions, there are important soil-plant relations that provide information about what the plant takes from the soil.

Soil-Plant Interactions

The current understanding is that there are fourteen "essential nutrients" that plants require for growth and successful reproduction.[89] Carbon, hydrogen, and oxygen are derived from the atmosphere and soil water. The remaining thirteen essential elements are supplied from soil minerals, soil organic matter, and organic or inorganic fertilizers. These are the nutrients that plants obtain from soils, and thus, are the potential source of soil fertility-derived differences in grapes and wines. Soil must be broken down into these essential nutrients before becoming available to the grapevine. Nutrients are taken up as dissolved ions (e.g., nitrate, aka NO^{3-}) from the soil solution, rather than from the soil particles directly (or from rocks). But the mineral nutrients have no established contribution to flavor. For any part of the plant to taste of the soil, the reconstitution of the soil within the plant would seem to be required.

Furthermore, the grapevine has some control over what gets in from the soil and up to the berries. Whatever is taken up must cross a series of cell membranes, whose purpose is to regulate traffic in and out of the cells. These membranes are the essence of biology, and are highly selective with regard to what is allowed to pass. The membranes actively exclude sodium and take up other essential nutrients. Soil minerals make it to the berries through one of the two systems of conduits in grapevines. Before ripening, most delivery is relatively direct via the xylem conduits running from the roots to all the shoot organs including the fruit. During ripening, there is no (or almost no) transport to the berry in the xylem; what nutrients that get to the berry during ripening pass through the leaves first, and are then delivered to the berry via the second system of conduits, the phloem (which run parallel to xylem conduits). Most flavors are then made in the berry.

It is possible that some flavor molecules might arrive in berries from the soil—assuming for the moment that there are flavors in the soil. Membranes are not absolute barriers; some herbicides, as well as excess boron, sodium, and magnesium, are taken up by roots in soils that are high in these nutrients. Getting a little bit of some molecule into the root from the soil or another plant may be expected. However, for this to be a molecule that is a flavor, accumulates in the berry sufficiently to be tasted, and remains stable through the fermentation into wines (at high-enough concentrations to be detectable by humans) is a stretch, given our current understanding of soil-plant relations.

If we accept for now that no flavor has been shown to come directly from the soil, then the effect of the soil is mediated through the biology of the grapevine. Unfortunately, the roles of soil mineral nutrition in grape and wine quality are not well studied, although the necessary nutrients for productivity are well known.[90] On the other hand, one of the clearest and most consistent soil-plant phenomena affecting grape ripening and the composition of fruit and wine involves vine water deficits in red winegrape varieties. Soil texture (the distribution of the sizes of soil particles) and soil depth combine to effect "water holding capacity," which in turn gives rise to a characteristic pattern of water loss from the grapevine between rainfall or irrigation events. Soil water holding capacity is one of the most important parameters of agricultural soils, and the same is true for vineyards. When vines producing red winegrapes experience (moderate) water deficits, more red color is produced, and often the fruit mature faster and have other sensory differences from wines made from vines that did not experience water deficits. The vine water status, however, is due to more than the soil; it is affected by rainfall or irrigation before and during the season and by the heat and dryness of the aerial environment, as well as by the size and spacing of the vines. The grapevine necessarily integrates its responses to its soil and aerial environments.

The effects of water deficits can be direct and indirect. For example, water deficits affect fruit metabolism directly, but also inhibit shoot growth. The effect on shoot growth may increase the fruit's exposure to sunlight in a way that affects fruit ripening, and comes through as a nuanced wine flavor. The direct effect in this hypothetical is the sunlight exposure of fruit.

Another indirect effect of soils could be realized in the fermentation products. Yeast growth and the progress of fermentation affect the formation of aroma compounds and are dependent on nutrition obtained from the fruit, although in some cases nutrients are added by winemakers. Volatile esters (aromas) are introduced into wine primarily by yeast during fermentation, although small quantities of esters are present in grapes prior to fermentation. Improved nitrogen status of fruit in turn can promote increased yeast growth and thus enhanced concentration of esters in the resultant wine.[91] Therefore, soil fertility can affect wines via its effects on the fruit itself, but also because the soil caused changes in the fruit that in turn caused changes in yeast growth and metabolism.[92]

There are additional ways in which the soil environment may provide or facilitate flavor. It has long been thought that plants transfer flavors to one another. Bioletti reported that the Roman naturalist Pliny the Elder knew what he was talking about when he said that grapes readily take up matter from surrounding plants, with wines tasting of the willows growing around the vines, and that it is well known that a couple of California weeds "communicate their flavor to the grapes." This belief was common enough to appear in Jullien's (1866) glossary:

> Gout de herbage. Il provient soit de plantes dont les racines, s'entrelaçant dans celles de la vigne, transmettent leur gout au raisin, soit des différents végétaux employés en remplacement du fumier.

> Taste of the pasture. It comes either from plants whose roots are woven into those of the vine, pass their taste to the grapes, or usage of various plants instead of manure.[93]

There has been evidence of water transfer between plants since at least 1957,[94] as well as recent evidence of nitrogen or phosphorous transfer via mychorrizal (fungus) connections between plants.[95] Thus, this kind of plant to grapevine transfer is possible, and may occur under some conditions, but there is not good evidence of it yet. At UC Davis, we were unsuccessful in our attempts to induce root grafting between grapevines.

A contemporary example is the belief that the presence of eucalyptus trees near a vineyard imparts their characteristic aroma to wines. The key compound in "eucalyptus aroma" is known, and has been found in wines. However, studies have suggested that the compound can be made in the berries themselves or can be obtained from eucalyptus leaves that come in with the harvested grapes.[96] The eucalyptus story is still in development.

Another possibility is that the mere presence of other plants causes the grapevine to react in a manner that gives rise to flavors. Plants' responses to their neighbors are increasingly under study, and allelopathy, the phenomenon in which a plant releases a compound that inhibits other plants' growth, could be a possibility. Allelopathy was thought to be discovered when it first made news in a study of a desert shrub that appeared on the cover of *Science* in 1964. That was later shown to *not* be an example of allelopathy, and although there is at least one clear example, the phenomenon remains a somewhat troubled research area.[97]

Jullien described a region's wines as "flat and having a taste of terroir, which [is] generally disliked by people who are not accustomed [to it]."[98] Wines develop bad smells or "off aroma" when certain microbial problems are present. Two of these are pertinent here because they could contribute to or be responsible for the older version of *goût de terroir*. The most widely experienced of these is probably *Brettanomyces*, a spoilage yeast in beer and wines that makes an aroma that has been called "horsey," "barnyard," "sweaty saddle"; in short, stinky. These are terms used in the same vein of the older, pejorative uses of *goût de terroir*. Although *Brettanomyces* yeasts are not typically found on the surface of grapes, they are common in the winery. When present, the aroma gives the impression of a sweaty saddle or manure that is often attributed to soil, but is actually a product of yeast metabolism. Other terms used to describe *Brettanomyces* are "earthy" and "leathery," qualities long associated almost exclusively with European wines. "Brett" (as it is commonly referred to) or other microbial issues may have been the source of much *goût de terroir*—an obvious possibility that is seldom mentioned. Burgundian wines used to have a reputation for a Brett kind of aroma. A 2000 study of phenol volatiles in Pinot noir wines in Burgundy detected Brett in 50 percent of wines tested, but it was not clear how many had enough growth of Brett to produce detectable aromas.[99]

For some wine writers, merchants, and consumers, this sensory characteristic remains integral to their sense of Old World or French wines, and as such is desirable. It has been observed, for example, that "currently, there is a silly obsession with the presence of *Brettanomyces,* a yeast said to create unpleasant off-aromas and flavors that are associated with this feral quality to which I refer."[100] It is also rumored that Robert Parker's palate has an affinity for a little Brett; and that wineries would make a batch with just that, explicitly for Parker's tasting. Most conform to it being a spoilage organism. A review article on the topic notes, "Good cellar hygiene is the first step in controlling infection of *Brettanomyces.*"[101]

Brett growth is sensitive to pH and temperature. Finished wines could be bottled with Brett, but the wine conditions prevent active Brett growth, and the associated "barnyard" aroma or "taste of terroir" may dissipate with aging or upon opening. Jullien and others sometimes refer to a *goût de terroir* that goes away with aging, leaving a good-tasting wine.

A second example of a possible microbial source of *goût de terroir* is the compound geosmin, which produces the characteristic smell of moist soil. A well-known bacterial contaminant in water, geosmin has been shown to be produced by soil-borne bacteria *Streptomyces spp.*[102] Geosmin has been found in Bordeaux wines; however, the authors tracked its presence in wine to bacterial growth on the grapes themselves, rather than coming from the soil.[103] This raises the possibility that wine tasters had been tasting "earthy" all along, but it was due to a bacterial contaminant on the fruit. However, in that study, the concentration of geosmin in the wines was usually low and near the suspected limit for human detection in wines. Bacteria or other microbes can gain access typically through wounds. Aromas or flavors from microbial growth, other than the standard wine yeast, can be argued to be positive or negative—another example of the cultural and subjective aspect of deciding what we like.

Despite the long list of impediments to direct flavors from the soil, soils do have profound impacts on grapevine growth and fruit development. Let's assume that one can consistently detect a nuance in wines made from vines on soils of a certain type, and not on soils of different types. Rather than a flavor in the soil, it is more likely that we would discover some regulatory molecule in the soil flora or organic matter that, once it arrives in the berry, alters berry metabolism. For example, geosmin is chemically related to the plant hormone ABA, which by definition influences many aspects of metabolism. Although not common, some volatile compounds in soil can play a regulatory role in plant development.[104] The soil does play a role in the hormones sent from roots to shoots, primarily cytokinin and abscisic acid, and by definition hormones have powerful effects on physiology. However, most nutrients and plant hormones are involved in many metabolic pathways, leaving direct metabolic connections of a nutrient supply to a specific flavor or aroma compound difficult to resolve.

ECONOMIC FORCES IN THE USE OF TERROIR

When we left our review of the use of terroir, it was still being invoked as a pejorative, up to about the turn of the twentieth century. The rise of the use of terroir in a contemporary sense can be traced to two periods in which the economic viability of French winegrowers was threatened. The first pressure arose out of the most unfortunate wine interaction between the United States and France: the import of the phylloxera aphid into France from the United States and the ensuing epidemic and decimation of vineyards. The second pressure was the development of competition for French dominance in the global wine hierarchy during the last quarter of the twentieth century, largely from Australia and the United States.

Post-phylloxera Resurgence in the Use of Terroir

From approximately 1850 to 1900, an epidemic of the soil pest phylloxera developed and wiped out many French vineyards, greatly reducing production. The demand for wine, however, did not decrease with the disrupted domestic supply, and alternatives were developed, including low-quality importations and frauds. After a bitter struggle for solutions to the phylloxera problem, French viticulture got back on its feet, choosing to go with traditional winegrape varieties grafted onto phylloxera-resistant rootstocks, mostly from American species of grapevines, rather than grapevine hybrids bred from traditional varieties and the resistant species. When domestic production revived, the recovery was not accompanied by a comparable reduction in alternative supplies. As a result, prices and growers' profitability fell sharply after the turn of the twentieth century, strengthening the power of merchants within the various commodity chains.

To overcome this situation, growers in different wine-producing areas used their political influence to establish controls over fraud and total production, and to develop regional promotional campaigns.[105] The emergence of terroir as an explanation for fine winegrowing can be traced to these honest efforts to better the livelihoods of peasant winegrowers, initially in Champagne and Burgundy, by sorting out fraud and then promoting their wines (and other products). The appellation system developed in earnest as a response to oversupply, as well as the low quality and fraudulent wines that existed in the post-phylloxera European wine markets, especially in France. The laws allowed local winegrower organizations to police fraud, and then to restrict the production area, which is clearly an economic advantage to the producers on the inside. The producers rightfully argue(d) that the rules help protect the consumer from the purchase of fraudulent wines.

The documentation of this is well researched and recorded, perhaps initially by Marion Demossier in Burgundy[106] and Robert Ulin in Bordeaux, and developed into some specific personalities recently by historian Philip Whalen. Demossier describes the promotional program as an example of "producing tradition":

> Burgundians developed new cultural strategies to market their wines during the inter-war years. Regional leaders, cultural intermediaries, and the wine industry collaborated to overcome overproduction, prohibition, and foreign as well as regional competition by exploiting the concept of terroir to develop a repertoire of popular festivals such as the Gastronomic Fair of Dijon, the Paulée of Meursault, Saint Vincent parades, an annual wine auction at the Hospice in Beaune and a Burgundian Pavilion at the 1937 World's Fair in Paris. These drew attention to the unique qualities of their wines and suggested how they might be best consumed. This aggressive marketing strategy was so successful that it became a model for French agricultural products promoted through the *système d'appelation d'origine controlée*. The unification of natural resources, historical memory, marketing strategies, and cultural performance resulted in an imaginative and enduring form of commercial regionalism.[107]

Whalen relates how an organized movement developed in the 1930s, with the purpose of promoting Burgundy, its wines, and its success:

> Gaston Roupnel, a Burgundian provincial folklorist, was especially important in efforts to link terroir and Burgundian regional culture through the figure of the modern rustic vigneron.[108]

> Gaston Roupnel described his wine as a "subtle emanation from the soil" replete with euphoric qualities responsible for the regional *joie de vivre*.[109]

And Whalen quotes from a Paris exhibition report:

> It is appropriate to remind visitors—or to reveal to them—that, by the riches of its soil and the glory of its wines, Burgundy is one of the classic fiefdoms of French cuisine.[110]

Historian Kolleen Guy described the similar story that developed, albeit with more vigor, in Champagne. In a review of her book, *When Champagne Became French*, Jerry Gough remarked:

The character of the region, in turn, was thought to be intimately tied to a quasi-mystical, quasi-scientific concept of terroir, an untranslatable term that was related not only to soil, climate, and geology but also to the "organic, animating spirit" that purportedly informed the personalities of those who inhabited the land and were subject to its influences. France was often portrayed by the geographers of the period as an organism composed of separate but mutually dependent regions whose specific characters were most tangibly manifest in the distinctive products of the soil.[111]

According to Guy, the government ratified this view when it initiated in Champagne what would eventually become the Appellation d'Origine Contrôlée (AOC) laws in 1908. Whalen writes that the success of the Burgundian promotional campaign was the impetus for the AOC legal system.[112] Regardless, the spatial designation for a wine label (accompanied by numerous regulatory mechanisms) has become commonplace in winegrowing and since spread to other commodities, such as cheese, which has its own myths of fabricated tradition.[113] During the 1930s, regional syndicates promoted the appellation system in France to official/legal status, and terms such as Grand Cru came into existence. These syndicates obtained the power to establish extensive rules for grape and wine production methodology, which must be met in order to use their "quality control label" via the AOC. Ulin summarizes an analysis of the development of "social capital" in Bordeaux:

> I have argued that the current status of southwest French wines follows conjointly from their political and economic history and a process of invention that links place and individual property with the authenticity and quality of wine. Moreover, the invention of the winegrowing past contributed to the naturalization of conditions and criteria that are fundamentally social and historical. This, in turn, has reinforced social distinctions between growers and provided elites with cultural capital that is widely regarded as authoritative.[114]

Demossier adds:

> By emphasizing the importance of the terroir and also of the wines and their owners or producers, reputation and tradition are constructed while helping to disguise the reality of social change.[115]

The growers and producers were able to achieve government intervention in order to control fraud and establish regional appellations or producer cooperatives, which effectively limited the production under various wine labels and helped them win back market power from merchants. In the process, all concerned capitalized on the value of having an attractive story that included the regional terroir explanation for distinctive wines.

In a 1962 French wine dictionary, the definition straddles both positive and negative connotations:

> Terroir (taste of): This taste corresponds to the characteristic taste given to the wine by the soil. Sometimes the taste of terroir which characterizes a wine is a key element which makes it particularly appreciated. But often also, it brings to the wine an unpleasant taste which can have been communicated by too much manure, by particular manure, by a contact too long with the "rapes" [pomace], or by something unknown.[116]

Similarly, from a 1970 dictionary, in which the unpleasant possibility no longer appears:

> Terroir. This word, which designates the land, the soil, has, when it is about wine, a particular meaning in the expression taste of terroir. This one designates a characteristic taste, particular, almost indescribable, that all the wines have when they come from certain kind of soils.[117]

Note that in addition to the attribution of a taste to the soil, another concept appears—that this putative taste is characteristic; that is, it is present and can be used as an identifier, a particularly appreciated identifier of a place. This is a belief that may be true, but has a clearer basis in reality in terms of its marketing potential than in the actual grapes. As unclear and unestablished as a "characteristic" taste from the soil was, terroir gets even fuzzier with the claim of ineffableness. The taste becomes an unassailable assertion; the kind of claim that one philosopher dismissed in this way: "What we can't say, we can't say, and we can't whistle it either."[118] This indescribability is reminiscent of Matt Kramer's comments on the inaccessibility of terroir at the beginning of this chapter. If you can't tell us what it is, how could we possibly evaluate whether it is there or not, let alone from whence it arises?

The Judgment of Paris and the Resurgence in the Use of Terroir

The second situation that correlates in time with the dramatic uptick in the use of terroir (as shown back in fig. 29) is the increase in international competition in the world of wine.[119] The sense of that competition is punctuated by a specific event in which the unthinkable happened: non-French wines were judged by French wine experts to be the best in a blind tasting.[120] In a now-famous 1976 event organized by British wine purveyor Steven Spurrier, top French and California wines were tasted in a head-to-head competition. The 1973 Stag's Leap Wine Cellars S.L.V. Cabernet Sauvignon was ranked above four famous Bordeaux reds, including the First Growths Château Mouton-Rothschild and Château Haut-Brion; the 1973 Chateau Montelena Chardonnay was ranked even more clearly above its French counterparts. *Time* magazine's Paris correspondent, George Taber, was on hand for the tasting and broke the news.[121]

It turns out that on the day the competition results were announced, the Château Lascombes in Margaux Bordeaux was hosting a group of leading U.S. winemakers. According to Joanne DePuy, who organized the winemakers' tour, "At the lunch, our French hosts welcomed our group, noted that we were trying hard and would one day make wonderful wine in the U.S., and that they would be happy to share their knowledge and centuries of expertise, etc."[122] The French hosts were unaware, but their American guests had heard the results of the tasting in Paris. When some in the French wine elite heard the news, they didn't take the results lightly or well. As the Stag's Leap Wine Cellars website says, "Less might have been made of the whole thing had the French tasters been other than top-notch and had they been less disdainful toward the California selections," or in some cases what they assumed were California wines. The French tasters, indeed all, were stunned when the names of the wines were revealed. The impact of the tasting for California wines was immediate, like a "vinous shot heard round the world," as the *Wall Street Journal* put it, catapulting California wines onto the world stage by illustrating that exceptional wines could come from somewhere other than the traditional apex of the winegrowing hierarchy: the French terroir.

Warren Winiarski, then winemaker at Stag's Leap, discovered that the wine competition results provoked anger, animosity, and bitterness among some in France. He wrote in a trade article in 1991, "Afterwards I received several letters from members of the French wine industry saying that the queer results of the 1976 tasting were a fluke. In essence, their letters argued that 'everyone knows' French wines are better than California wines 'in principle' and always will be."[123]

Although the 1976 Paris tasting is the most famous, it is not an isolated case.[124] Similar results have occurred in several subsequent blind tastings that pitted top California and French wines against one another in competition. The French food and wine magazine *GaultMillau* staged a follow-up "Wine Olympics" in 1979, a huge affair with more than 300 wines from thirty-three countries. A higher number of French wines were entered, but again the California wines fared well. In 1986, a similarly impressive French and American expert panel placed Clos Du Val and Ridge, both California Cabernet Sauvignons, above the French Montrose, Leoville, and Mouton. The Paris tasting was also recreated at the Smithsonian's National Museum of American History in 1996 to mark the event's twentieth anniversary. At that time, bottles of the original first-place wines—the 1973 Stag's Leap Wine Cellars Cabernet Sauvignon, and the 1973 Chateau Montelena Chardonnay—were accepted into the museum's collections. Results similar to the first Judgment of Paris were obtained, meaning that there was little to separate the top wines, and it was certainly not country of origin.[125]

At an International Pinot Noir Celebration in Oregon, seasoned Burgundy winemakers often couldn't distinctly identify the Burgundian Pinot noirs from the Oregonian wines. Similarly, the winemaker for Tattinger Champagne noted the inability of most tasters to differentiate between French champagnes and their counterparts produced in California, and this can be expected in some comparisons of wines from other regions. One more anecdote: in reviewing *The Accidental Connoisseur,* the economist Richard Quandt says that the author found wine experts who say absurdities like "Burgundies are the greatest wines to drink. . . . California is a joke by comparison";[126] Quandt refers to his own contrasting data: "In a recent [2004] tasting of ours, $100-$150 California Pinot noirs beat Burgundies in the $300-$600 range hands down."[127] That a California or Oregon wine was ranked at the top of some tastings is less important than the lack of clear differences among top wines based on the country of origin.

The difficulty in identifying the origin of top wines in blind tastings suggests that, given sufficiently similar environments, genotypes, cultural practices, and winemaking technologies, similar (but never the same) wines can be produced from different locations—if that is one's objective.[128] Furthermore, even if there are important flavors of the wine that derive from the soil, it is possible that with sufficient understanding of soil-plant relationships, the soil (or other components of the production system) may be managed toward those desirable flavors, perhaps even to the extent that, rather than ensuring something unique, the wine from one soil could be made *indistinguishable* from the wine from another soil. This logic applies to an expanded version of terroir that includes any number of other aspects of the environment as well.

The similar results of repeated Judgment of Paris–type experiments demonstrate that similar wines can hail from various places. It is also possible that similar wines may be produced by more than one permutation of the many factors involved. As skill in winegrowing improves, more similar wines can be expected when winegrowers adhere to the same wine model, although each lot of grapes and each fermentation is different. New Zealand viticulturist Michael Trought points out that the California wines were the same before and after the Judgment of Paris; all that changed were some people's expectations as to how the wines may taste, and the appearance of a chink in the perceived armor that is the primacy of French wines.[129]

The Traditional Hierarchy

We've noted that several of the top estates had their reputations codified by the mid-nineteenth century. Acknowledging this, French enologist Émile Peynaud noted that wines of the Middle Ages were sold and drunk young because they would not keep, and said of wine of the 1800s, "Although they made the reputation of our [French] appellations, [they] were green, astringent, low in alcohol, and kept badly."[130] The formalization of the hierarchy of classed growths in Bordeaux (First and Second Growths, etc.) was not based on the terroir explanation, which according to French wine historian Philippe Roudie, was originally not even part of the Bordeaux vocabulary. Rather, the terroir tradition was adopted from parts north after it had proven successful. Wine writer Frank Prial gave a somewhat tongue-in-cheek description of the famous 1855 classification:

> What happened in 1855 was this: An international exposition was held in Paris to try to top the Great Exhibition held in the Crystal Palace in London four years earlier. The organizers, apparently as confused about Bordeaux wines then as many of us are now, asked the Bordeaux Chamber of Commerce, in effect: Of the thousands of wines you've got down there, which are the best, anyway? The chamber passed the query on to the Bordeaux wine brokers, who replied, in effect: Good question. We'll put together some kind of a list.[131]

The list was explicitly based on price, not terroir or wine taste, although the price probably corresponded with the reputation of the producer.

What wine lovers consider good and seek out has not been stable over time, being subject to both improvements in the vineyard and the winery, and marketing manipulations.[132] Today we do not accept green, harsh wines that don't keep. Since the development of the traditional hierarchies, winegrowing has changed dramatically, including spacing, trellising, training, preplant soil preparation, analysis for (and utilization of) fertilizers, and pest and disease control measures, in addition to largely favorable climate changes. Among the known consequences are the increased ability to ripen fruit, consistently obtain higher yields, and regulate the cluster microclimate. The most basic consideration in the wines, the alcohol concentration, has increased more than 50 percent since the 1800s, a tribute mostly to improved vineyard management.

In many cases, even the winegrape varieties grown in higher-classed vineyards are different now than when the original reputations and high prices were established. Cabernet Sauvignon is much more dominant today; once common Malbec, Carmenere, and Petit Verdot have largely been replaced. It also used to be standard to mix red and white grapes in the fermentation (not that that was a mistaken technique by any means). Today, virtually all of the roots—the part of the plant that interacts directly with the soil—are of completely different plant species than those grown prior to the phylloxera devastation of the late 1800s.

The changes in winemaking technology since the establishment of Old World reputations are too many to list, but these advances should be noted in general for their large impact on fermentations, wine style, and wine health (i.e., the absence of spoilage). The result is that in addition to changes in the ability to grow grapes and make wines, wine styles (or wine models) have changed dramatically. The wines that are made and appreciated today by an increasingly knowledgeable and sophisticated market are not the wines upon which the early reputations were established.

Despite all this, the same First Growths remain at the top of the wine world hierarchy. Early on, it may have been that the best wines arose from these leading producers, in large part because they had the wealth to invest in the vineyard and winery technology. Regardless, the wine laws holding the hierarchy in place have also restricted opportunities to change.[133] In much of France and Germany, the same basic hierarchy is in place today as 150 years ago or more. There must have long been agents interested in change, but the hierarchy "turned what was supposed to be a consensual truth into an inconvenient truce."[134]

In general, the laws that affect supply have long been influenced by the high-end producers. This description by eminent economists Giulia Meloni and Johan Swinnen is representative of recently published evaluations of the development of European wine laws:

> It is clear that some of the EU wine quality regulations have strong income distribution effects because they require, for example, access to specific plots of land in specific regions. In fact, the official EU regulations explicitly

specify that "the concept of quality wines in the Community is based . . . on the specific characteristics attributable to the wine's geographical origin" (OJ 6.6.2008: Council Regulation No 479/2008 of 29 April 2008, Article 28). This approach to quality regulation in the EU is not new or unusual in wine-growing. Throughout history, quality regulations for wine have been motivated both by efficiency considerations and in order to restrict the production of wines to certain regions (which increased rents for land- and vineyard-owners in those regions). Owners of vineyards and wine producers have been among the rich and powerful. Not surprisingly the profits and power of existing wine producers and vineyard owners attracted others to invest in wine production and induced innovations. These new investments and innovations threatened the rents and power of the established vineyard owners and caused protectionist reactions. There are many examples of such political economy processes which resulted in significant regulations in wine markets during Roman times, the Middle Ages, the Renaissance period and in the past few centuries. Whenever changes threatened to reduce their rents, established producers have sought to constrain or outright remove the threat of new developments through political means. They lobbied governments to constrain threats to their rents through regulatory initiatives. Because of their wealth and power they were often successful.[135]

The hierarchy's lasting power in the face of changing production systems and wine models makes more sense if considered as a skill in controlling the market, rather than having a monopoly on special environments. Although the promoters of the terroir explanation frequently cite Roger Dion's historical research on winegrowing in France, it has not been in a way that carried the sense of his viewpoint on the attribution to soils or a physical patrimony. The preface of his 1959 *History of the Vine and Wine in France* makes clear that

> Roger Dion himself, incontestably the greatest French historian of wine, wondered consistently throughout his writings why French producers preferred to agree that the qualities of their wines were the effect of natural privilege, or as he said, "of a particular grace accorded to the soil of France, as though our country would derive greater honor by receiving from Heaven, rather than from the painstaking labors of man, this fame for the wines of France."[136]

For nearly 250 years, terroir was construed as a certain kind of soil giving rise to a certain kind of taste. It turns out that the wine laws of Europe have been acknowledging, step by step, that the soil-based or broader physically based claims have not been substantiated (not that that is impossible, but it just has not been done). According to Dev Gangjee, professor of intellectual property law at the University of Oxford, European countries have spent decades gradually distancing themselves from a system in which physical geography has greatest importance. He says that French wine law has gone further: "A historical perspective therefore allows us to appreciate the French AO regime's eventual rejection of '*conditions particulières de climat et de terroir*' as the sole or sufficient basis for protection, while also exposing an alternative basis founded on intergenerational, collectively generated *savoir faire*."[137] Accordingly, economists Tim Josling (in his 2006 Presidential Address to the Agricultural Economics Society)[138] and Tomer Broude[139] tell us that arguments from the EU in trade talks moved in the direction of continued terroir protections (Geographical Indications) for wine and other "widgets" on the basis of special *cultural* products.

Renowned New Zealand geographer Warren Moran devoted a significant part of his career to studying terroir and appellations. He concludes that for quality wines in Europe, and France especially, the outcome is clearly advantageous for those who hold the monopoly:

> Appellation systems are an example of an external economy imposed by a group of producers who wish to maintain the benefits of their real or created advantage and make it more difficult for other producers to benefit. In differentiating a product by its area of origin they are creating a type of differential rent . . . which may also be viewed as a type of monopoly rent. . . . By restricting supply and putting barriers on entry to the production they are likely to increase the price of the commodity that they are differentiating.[140]

The terroir concept, although perceived and promoted as legacy by some, centers on an active and conscious social construction, which shapes current place-based identity and emphasis.[141] (See also the Ulin and Demmosier studies cited earlier in this chapter.) In his study of Appellation Cassis, the third oldest and one of the smallest appellations in France, geographer Daniel Gade refers to a process of "patrimonialization" used as a counterforce to the homogenizing trends in the globalization of world food systems (in the minds of many toward the late-harvested "fruit bombs" of the previous chapter), and observes:

> In [the AOC concept], the characteristics of a place—the terroir—are used to gloss its legally protected, territorial definition on which hinge claims to a place-based product authenticity and, by extension, quality. . . . Wine types have evolved in spite of the absence of real innovation; political territory has been used to define terroir; the discourse of quality depends heavily on the historic past; vineyards have acquired a community value beyond any productivity; and producers have defined and defended their territory to boost its prestige to themselves and their consumers. The key entity in the appellation is the lower winegrowers syndicate. Presumptive statements, promotional rhetoric, consumer desire, and the politics of local decision making have shaped this wine region far beyond its environmental associations.[142]

A common denominator among the terroir systems is the belief that wine production and their labels must be limited in area and production. According to agricultural economists Rolf Mueller and Dan Sumner, the most important characteristic (economically) of a terroir wine is the protection it is afforded by state government.[143] The wine appellation is used in a strategic role to make it appear as if the control of territory also controls cause and effect—controlling the area for the wine label looks like the physical boundary is the cause behind the fineness and rarity of the wine, but according to most investigators, the boundaries of the territories are set by economic and cultural agendas more than aspects of the environment that are known to cause wine flavors. In the AOC system, with all its controls over production, when one says, "We have a good terroir here," it is like saying, "We have good rules here," which lead to good products.

Although the terroir marketing approach has spread to other wine-growing countries and products, with more than 350 appellations in France, some consider the system unwieldy for consumers. Historian Harry Paul, author of *Science, Vine, and Wine in Modern France,* called it "a sick system."[144] The more recent laws, designed ostensibly to protect consumers from bad wine in the EU, prevent wine made in the EU from being called wine if it is made from imported grapes.[145] In Italy, where the rebellious SuperTuscans challenged the system and made successful wines outside the regulations, authorities changed the regulations to fit their innovative methods.[146]

ON SHAKY GROUND—THE DIFFUSION OF TERROIR INTO SCIENCE

It would be encouraging, even gratifying, if terroir's broad incorporation into the world of wine were based on sound evidence connecting the grapevine to grapes and wines, and resulted in a consensus definition. Unfortunately, the "discovery" of terroir in the popular press was not preceded by scientific discoveries of soil-derived flavors, or other validations of putative characteristic flavors from a more broadly defined terroir.

The transformation of terroir in winegrowing from a pejorative to laudatory descriptor coincided with the emergence of the appellation system as a legal and marketing institution, and the surge in terroir use in English corresponds loosely to the Judgment of Paris and new international competition in the wine world (fig. 29). Terroir emerged in the academic literature at approximately the same time (fig. 34), and therefore

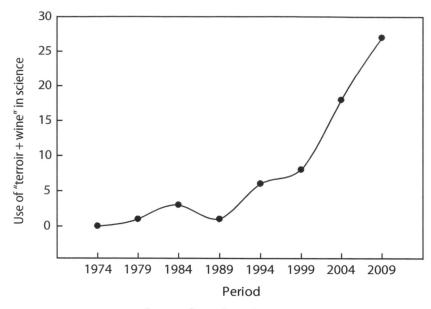

FIGURE 34. FREQUENCY OF "TERROIR" AND "WINE" APPEARING TOGETHER IN RESEARCH PAPERS IN THE CAB ABSTRACTS DATABASE. The initial movement of "terroir" into scientific or quasi-scientific publications followed the 1976 Judgment of Paris tasting by a few years, and the larger increase commenced in the 1990s.

may be part of the same international wine competition and labeling squabble. It turns out that a new French word for characteristic taste of wines from a place, *typicité,* was adopted (if not coined) in 1979.[147] The frequency with which typicité occurs in Google digitized books closely corresponds to the increase in terroir use (fig. 29). Perhaps these are all coincidences, but the observations fit the argument that after the Judgment of Paris, the world of wine got busy building a defense against the real or imagined New World threat to the Old World wine hierarchy.

By the early 2000s, typicité was similarly employed in English, and some students arrived in my viticulture classes with typicité firmly cemented in their vocabulary.[148] The wines from a certain terroir (and a label indicating same) are now supposed to have a specific style and a characteristic taste, a "distinctive taste deriving from the growing conditions."[149] According to the description of terroir in *The Oxford Companion to Wine,* "Over those centuries the special character of each vineyard has emerged, either as reality or, just as commercially significant, accepted folklore."[150] Typicité is the terroir myth all over again.[151]

Academic publications using terroir include the works of economists and geographers, some of whom were cited earlier in this chapter, as well as the works of wine and vine scholars. The viticulturists, who are trained in plant sciences, are the last and least likely to adopt terroir.

The scholarly work in viticulture in which terroir appears has taken two main directions: elucidating zoning criteria using soil and other environmental characteristics for the establishment of terroirs for inclusion on wine labels; and associating differences in fruit or wines with possible causes in the grapevine environment. The former, which is essentially terroir defining, has the difficulty of needing the latter (that is, knowledge of which environmental parameters lead to which distinct flavors) on which to build a legitimate basis for zoning, and as a result is usually subject to the criticism of putting the cart before the horse.

Early terroir papers appeared in the quasi-scientific journal the *OIV Bulletin* with titles such as "Importance of Terroir as a Factor Differentiating the Quality of Wines" (1984) and "The 'Terroir', an Indispensable Concept for the Establishment and Protection of 'Appellations d'Origine'and to the Management of Vineyards: The Case of France" (1987). In his 1994 paper "What Is Terroir? . . . ," discussed earlier in this chapter, Falcetti lists a menagerie of terroir definitions and then identifies approaches to defining terroir zones for winegrowing. These opinion pieces defend terroir's legitimacy or even assert its undeniability against an unclear challenge—perhaps in reaction to the negotiations over labeling laws, referred to by some as a "war on terroir."[152] Consistent with that suggestion, the zoning group introduced additional aspects—natural *and* immutable—to the meaning of terroir:

> An essential notion is that all its components are "natural", and cannot be significantly influenced by management.

> The term "terroir" is used to represent the group of environmental factors (climate, relief and soil/subsoil) which are important for wine quality but which cannot be modified by the grower for technical/economic reasons.

> A terroir, therefore, is defined as a complex of natural environmental factors that cannot be easily modified by the producer. With the aid of various management decisions, this complex is expressed in the final product, resulting in distinctive wines with an identifiable origin. The terroir cannot be viewed in isolation from management and cultivation practices, although they do not form part of the intrinsic definition.[153]

In this case, "natural" evidently means without and/or unspoiled by human intervention.[154] This is a social or cultural issue, not a path or condition for understanding how the environment affects fruit and wines. Agriculture is by definition human intervention, an intervention that allows us to sit down and enjoy a glass of wine rather than perpetually chase food. Eminent French enologist Émile Peynaud rejected the natural/interventionist attitudes in winegrowing, pointing out that human intervention is everywhere in the process, both historically and currently.[155] Moreover, the grapevine's responses to the environment are not predicated on whether the environment became that way by human action; the grapevine has no way of evaluating that. For example, the grapevine does not distinguish potassium presented in fertilizer from that which is freed from the native soil minerals. There are, no doubt, unchanging characteristics of vineyards, such as latitude, distance from the ocean, and so on, including the parent material of the soil, and some of these *can* be important for vine growth and fruit ripening. However, a phenotypic trait in the grape does not depend on the immutability of environmental attributes. Again, the berry has no sense of whether its exposure to light is a function of its latitude or whether the water supply in the soil is a result of the climate or from irrigation.

The immutable aspect of the environment can be used to define geographical space, but in the terroir explanation it is expected to contribute a distinctive or unique flavor. We saw where this flavor became an explicit aspect of the positive terroir definitions that emerged in the 1970s. Every place on Earth can be considered unique. For that matter, every grapevine, every cluster, and every berry is unique. Yet each cannot be counted on to make a wine with a unique flavor. Many of us would love to have a vineyard that produced uniquely wonderful wine flavors, and that fortunate situation is the presumption around which territorial lines are sought. These approaches and underlying notions are each consistent with a belief that the supply of grapes and wine must be limited by some physical attribute of the environment that cannot be reproduced elsewhere—which would, coincidentally, produce an economic or marketing advantage.

The other direction of viticulture research in which terroir is invoked makes contributions by correlating measurements of some environmental parameter with some attribute of grapes or the resultant wines. This research can answer questions about the impact of yield, soil fertility, and temperature during ripening on the resultant grapes and wine. Studies with a couple of varieties, a couple of seasons, and a couple of soils are standard field research agriculture. What distinguishes these works from other agronomic and horticulture research is their use of "terroir." In most cases, the terroir-adopting scientists tell us that they are studying terroir in the introductions and conclusions of studies, but the new knowledge is in the results showing, for example, that clusters that experience greater early-season light intensity have lower MIBP in grapes and reduced vegginess in wine aromas. The authors could just as well have reported the data, and described how fruit composition did or did not vary, without mentioning terroir, as in this 2005 report on the composition of berries from different vineyards: "This study showed that the combination of 1H NMR spectra with chemometric methods by multivariate statistical analysis is able to discriminate between berry samples from different environments or terroirs."[156] This explains how terroir has remained outside the scientific literature, despite being around wine for centuries and despite viticulture research institutions operating in all major winegrowing countries. From a biological standpoint, "terroir" is a kind of window dressing at the beginning and end of scientific studies of the grapevine. The term *terroir* is by and large not used in leading plant journals, there are essentially no scholarly works invoking the terroir of plants besides grapes, and there is no evidence of grapevines having unique reactions to environments. *Terroir* appears in less than 1 percent of the instances in which science journal papers address "grape + wine," and the papers that do use the term have not been highly cited. Thus, the scientists who employ *terroir* are firmly in the minority, if not on the fringe. Agriculturists do not use terroir in investigating crop development, because that development is understood and studied in the context of Klebs' Concept—wherein the interactions of the variety with its environment cause physiological changes that we realize as different phenotypes (read: grapes). Invoking "terroir" in grapevine biology is largely gratuitous, because the term is at best a synonym for *environment*, used as jargon in studies of grapes and wine.

Although the popular and scientific literature gives the appearance of a consensus around a broadened sense of terroir that extends to the synonym for *environment*, soil remains a more important aspect for traditionalists. Still in transition in English, *goût de terroir* morphed from a soil-derived off flavor to what was described in 1930 as an earthy and undesirable taste to a more neutral earthy taste derived from the environment in 1967,[157] but was again a soil-derived flavor when used as an honorific in 1979 in a *New York Times* wine description: "The Rouchottes in particular was a classic Burgundy, with color, body, and bouquet, and the unmistakable *gout de terroir*, the taste of the soil that distinguishes the great wines of the Cote d'Or."[158] Most wine experts and scholars acknowledge the role of the weather and so on; however, it is telling that soil is often referred to separately from the environment, as in David Skalli's allusion, cited at the beginning of the chapter: "nearly perfect fruit, climate, and terroir." Accordingly, a 1995 wine dictionary entry for *terroir* reads: "Terroir: combination of soil, subsoil, exposition, and environment which give the wine its own characteristics."[159]

My experience (and this comment is based on my experience alone) is that even though it is common to see terroir described as inclusive of environmental factors, for many wine experts, soil remains predominate with a somewhat reluctant acquiescence to nonsoil factors in wine flavor, and discussions are inevitably brought back to soil. Falcetti identified "bioclimatic" and "geo-pedological" approaches to terroir zoning. The bioclimatic approach is criticized for not having a soil component; however, the geo-pedological approach is *not* criticized for its omission of weather or climate.[160] In the 2000 *Handbook of Enology*, terroir is still defined as soil and explicitly not the aerial environment: "Terroir, a term coined by the French, refers to the influence of non-climatic environmental factors (soil, topography) on wine composition and quality."[161] Not surprisingly, many terroir discussions are about the meaning of the word, rather than specific vineyard conditions and consequences.

I should address one additional aspect of the diffusion of terroir into science. In both directions of viticulture research on terroir, some experts include winemaking in their definition of terroir and add the biology of what makes a good or distinctive winegrape to the ongoing Old World–New World skirmish over what to put on a wine label.[162] The following are a few examples.

> Two different perspectives have emerged in worldwide wine production to indicate the origin and quality of grapes and wines. The first, common in the New World countries, indicates the grape cultivar on the label. The second, common in the Old World countries, is called "terroir" and is used to describe all aspects of the environment, which include the soil, climate, and cultural practices.[163]

> Terroir has been acknowledged as an important factor in wine quality and style, particularly in European vineyards. . . . It can be defined as an interactive ecosystem, in a given place, including climate, soil, and the vine (rootstock and cultivar). Some authors also include human factors such as viticulture and enological techniques in their definition of terroir.[164]

> The underlying concept of "terroir" includes all regional parameters with an impact on wine composition such as soil, climate, viticulture management, crop level, and wine-making procedures. In Burgundy, the quality hierarchy of wine is established on a classification of vineyard sites only, while in Bordeaux,

individual wine estates are classified, combining the "impact of terroir" with the various effects due to individual viticulture and enologic practices.[165]

Sometimes the act of demonstrating that soils or the environment matter to grapes and wines is used as an exercise in validating the idea of terroir. The authors of the third quotation above found that wines differed a great deal depending on who made them, even if they were from grapes with the same "vineyard designation." A group of producers in Oregon recognized the role of winemaking was potentially important, and initiated their own study. The International Pinot Noir Celebration funded several experiments making traditional small lots from three vineyards. Each lot was then divided into three, with each sublot vinified at one of the three wineries. In a blind tasting (and chemical analysis) of the nine different wines at the 1998 event, "the hand of the winemaker clearly stood out to identify the wines in terms of style rather than earth."[166] In a similar experiment in the 2000s, the sommeliers could not distinguish correctly in 60 percent of the trials between wines that differed in the grapes and wines that differed in the winemaking. A third version of the exercise also found the winemaking to be clearly predominate in its effects on wine sensory attributes.[167]

Winemaking affects wine. This isn't news, nor should it be news when a viticulture study finds "a significant influence of soil and climate on the grape composition."[168] When terroir is used to include all factors that can affect grapes and wines, it only means that wines are different. If wines did not differ in a systematic way, wine would not be interesting or worthy of reflection, study, or argument. There is no question that wines from different winemaking practices conducted on the same grapes are different, or that grapes from sufficiently different environments using the same winemaking differ as well.[169] The challenge for the viticulture side of winegrowing is to learn which parts of the environment impact vine growth and development sufficiently to result in significantly different fruit, and to exploit that knowledge to identify the best sites and practices for selected wine models.

CONCLUSIONS

Although the grape is necessarily a product of its environment, there was a premature effort to place a natural and scientific imprimatur on an economic instrument. The consensus of scholars of a wide range of disciplinary approaches is that the general attribution of wine quality to the physical environment of the terroirs defined in appellations, and to soils in particular, has more social or economic basis than empirical evidence in the grapes and wines. The increase in the use of terroir is a cultural (and economic) curiosity, rather than an agronomic phenomenon. Agriculturists do not employ *terroir*, and no discoveries led to its wide adoption. The impulse in the world of winegrowing to attribute wine qualities to an inherited physical geography (most often the soil) as patrimony that is to be protected was acted upon without doing the work to discover how (or even if) terroir is the causal factor.

While soil and plant scientists seek insights into the environmental parameters that can be traced in the fruit and wines, *terroir* has remained too nebulous to be useful in any genuine inquiry into what is inside the grape or the bottle. The term appears in a few scientific papers, many of which are defenses of a traditional explanation of zoning and wine labeling. Thus, terroir has no place in the ecophysiology of the vineyard (or of any other plant) and makes only an ignominious appearance in scientific literature. In *Terroir*, James Wilson says "good and great wine can only come from certain very limited environments—the terroirs." Once again, "terroir" is superfluous to "environment." By definition, the best is a limited fraction of the total, and what insights saying "terroir" offers into the "best" grapes and wines remains a question. A genuine inquiry about the nature of the winegrape can omit terroir and still consider vineyard factors affecting grapevines. Give it a try.

Terroir reflects a wine business perspective that is manifested on wine labels. Today, *terroir* is primarily a marketing term that mixes extrinsic and intrinsic wine properties; when formalized into laws restricting production, it becomes a system to extract high rents. The use of *terroir* has expanded to cheese, coffee, and other products because of its success in selling products, just as it expanded from Champagne and Burgundy to the wine world at large.

Regardless of any controversy over the appellation system and wine labels, the French have made what have been universally recognized as some of the world's finest reds, whites, dessert, and sparkling wines for hundreds of years. Accordingly, French wines have been the models for many a winemaker, probably nowhere more so than in New World winegrowing. In accepting an award from the American Society for Enology and Viticulture, Justin Meyer, who established the well-respected Silver Oak Cellars in Napa, said he simply tried to identify what the French do, and then did that.

In addition to the hundreds of years of winegrowing tradition and premier wine reputations, France produces about twice as much wine as the United States, and the French drink about five times as much wine per capita. Historians concur on the assimilation of wine and terroir into the French ethos. In *When Champagne Became French,* Kolleen Guy describes the development of terroir into a kind of French identity concept, along with the initiation of the legal boundary and labeling for wine based on terroir in Champagne.[170] French historian Georges Durand says the same of wine, calling it the "distinctive national reference,"[171] and French philosopher Roland Barthes offered a similar assessment, that "wine is felt by the French nation as national property."[172] French winegrowers have earned the right to wax on wine. Much of the non-French wine world (again, perhaps the New World in particular) has sought out not just what the French and other respected Old World producers have done to make great wines, but also what they have said, and then repeated it themselves.

English speakers who looked to the Old World traditions may have naively assumed that since there was a word that meant a characteristic taste given to the wine by the soil (or broader environment, or environment + cultural practices + winemaking), that terroir was established fact. However, the Old World traditions that New World producers have sought to emulate include success in the whole wine enterprise, where part of the success lies in marketing and the traditional explanations.

Terroir is patronizingly described by promoters outside of viticulture as complex and difficult to grasp. In *The Oxford Companion to Wine,* terroir is a "much-discussed term for the total natural environment of any viticulture site. No precise English equivalent exists for the quintessentially French term and concept." My difficulty is not that terroir is of French origin, complex, or difficult to translate, but that its use defies a clear understanding. Trying to resolve the role of terroir in winegrowing calls to mind Paul Gross and Norman Levitt's description of postmodernism: "To give a concise statement of the postmodern doctrine would be an almost impossible task . . . for postmodernism is more a matter of attitude and emotional tonality than of rigorous axiomatics."[173] One need only replace "postmodern" and "postmodernism" with "terroir." When I was beginning to investigate terroir several years ago, I was standing at the photocopy machine (with yet another definition of terroir) when Professor Boursiquot from Montpellier, France, who was visiting UC Davis, happened to pass by. I commented on how I was collecting definitions of terroir, and he replied, "That would make a good hobby." In essence, one needs to have conversation about what terroir is before having a discussion about its role in winegrowing.

The lack of clarity may not be entirely accidental. For example, *Newsweek* journalists tell us that "even the greatest wines—maybe, especially those—have qualities that can't be measured in a mass spectrometer. Terroir is best expressed by metaphor, not chemical analysis."[174] Madame Lalou Bize-Leroy remarks in the same vein, as quoted in James Wilson's *Terroir,*

> Suppose that one day we might know the nature of all the elements which compose this nourishing "new earth" between the pebbly surface and the Jurassic sub-basement, would we then be able to determine exactly the physical and chemical influence of these elements on the intrinsic, fundamental character of the wine issued from them? Personally I do not believe so, for all of this is too dynamic—I would say almost alive—to put in an equation.[175]

In a revealing article in *Discover* magazine, Robert Kunzig reported that when French soil scientist René Morlat offered to conduct standard soil analyses in some of the best Burgundy vineyards, "the response of the vintners was layers and layers of ripe red raspberry."[176] In the same article, Michael Feuillat, director of the Institute of Vine and Wine—Jules Guyot at the Université de Bourgogne, stated that "Burgundians aren't too hot on having a terroir study here like the one they did at Angers. . . . They say, 'You're going to demystify everything. If you start saying a grand cru means such and such a percentage of clay and limestone, such and such a slope and nutrition of the vine—it will lose all its sacredness.'"[177]

James Wilson says that terroir has a spiritual aspect. In reference to doubters he laments, "I puzzle at their real appreciation of wine." Thus, according to Wilson, in order to appreciate wine, you must join believers in the chant of "terroir." Following up on Kramer's definition of terroir (presented at the beginning of the chapter), terroir is not only of another dimension; it takes on human characteristics: it sanctions, prospects, and "in this it is at odds with science, which demands proof by replication rather than in a shining uniqueness. Terroir is the more beautiful question."[178] This is poetry, and a good example of terroir written "with their most lyrical pen," as UC Berkeley sociologist Marion Fourcade recently and so fittingly put it.[179] I think Matt Kramer is a gifted writer, but these descriptions sound like cautionary "You can't get there from here" warnings for those seeking clarity about terroir and the nature of the fine winegrape.

Then again, not getting anywhere may be the point. As I continued to pull together opinions on terroir, the collection of definitions developed a historical trajectory, but not toward a more coherent understanding of how the grapevine and its environment interact to make a grape. Terroir was transformed from bad to good with no basis in the grapevine, and from a literal soil flavor to a tardy and inconsistent acknowledgment of the aerial environment (by some authors), and then went wobbly with ineffable flavors and shining questions. In the end, terroir is a shibboleth that establishes an in-group in a world unto itself. This isn't wine appreciation, and it certainly doesn't reflect interest in the grapevine; it is more like wine snobbery. After investing greatly in trying to sort out what terroir means for grapevines, I suggest following wine writer Paul Lukacs's advice, and not get hung up on it.[180]

"I like to think we aren't so much anti-science as we are pro-myth."

Christopher Weyant/The New Yorker Collection/The Cartoon Bank.

Epilogue

In the beginner's mind there are many possibilities, but in the expert's there are few.

—Suzuki Roshi

Wine is a traditional product with traditional explanations. Today more than ever, the traditional explanations for wine quality and the stories that accompany the wines we drink are cherished almost as much as the wine itself. In this work, my goal was to explore how the traditional explanations for the vineyard origins of fine wine have gained wide acceptance, and to compare the expectations raised in each to the empirical evidence. My analysis exposes the traditional explanations in winegrowing to be a far cry from the settled issues they are portrayed to be.

It is not surprising that early growers and producers offered explanations for grape and wine quality—after all, it is human nature to make up stories to explain our observations. The terroir explanation and the concepts of High Yield and Low Quality (HYLQ), Big Bad Berry (BBB), vine balance, critical ripening, and physiological maturity have each contributed to the contemporary zeitgeist of winegrowing. These winegrowing myths share a similar background as simple and intuitive concepts that ascended to prominence with little investigation beyond casual observations and anecdote. Although sometimes vague, the myths of winegrowing make predictions about the grapevine that can be approached relatively easily. What we know about grapevines from the empirical evidence is consistent with some of the received knowledge, but the evidence also refutes these myths or at least finds them wanting in many regards. Having gone unexamined and untested for decades, the major myths of winegrowing (the soil as the origin of flavor, the dilution of good flavor by increasing yield or berry size, the balance of leaves and fruit as a key determinant of winegrape quality, and the critical nature of late ripening and the harvest decision) seem to have developed into beliefs, sometimes passionately held, yet based largely on intuition and the power and influence of expert opinions.

MYTHS AS RECEIVED WISDOM

Much of this book has involved reviewing empirical evidence from grapevines that runs counter to the popular myths of winegrowing. It turns out that, except for the concept of vine balance, the evidence does not support these myths or explain why they became the paradigms that they are today. Even the support for vine balance (a concept developed by viticulturists) is suspicious, if not misguided.

Let's briefly review some of the difficulties that surfaced regarding the popular winegrowing concepts examined in this book. In my review of the scientific literature, it was surprising to learn that numerous studies from all over the world (including historical, regional, and informal contemporary experiments, including vineyards planted in premium winegrowing regions with many of the premier winegrape varieties) have reported little or no response of fruit and wine to changes in yield.[1] When the consequences of establishing low yield or small berries in fruit and wines are inconsistent and sometimes contradictory, rather than falling into line with HYLQ and Big Bad Berry predictions, one should look elsewhere for fundamental truths about the grapevine. Because quality attributes of the fruit and their wines depend on the ways in which yield and berry size are changed, the straightforward implication is that it is the developmental journey (i.e., the environmental conditions) during the season, not the final yield or size destination per se, that is more important in determining winegrape quality.

The concept of vine balance, usually evaluated as yield:pruning weight (Y:PW), was developed by viticulturists as a solution to their inherent skepticism of HYLQ and recognition that observations were often inconsistent with HYLQ. However, in many cases where yield is directly manipulated in healthy productive vineyards, yield and Y:PW are very closely correlated; therefore, Y:PW is unlikely to provide greater insight than yield itself. The balance concept, as a key to winegrape quality, has a dubious physiological basis with respect to identifying a balance point that produces optimal fruit composition, and there is little more than faith supporting the vine balance paradigm for high fruit quality as normally expressed in a Y:PW range of 5 to 10. Unfortunately, viticulturists moved on to balance from yield as much by fiat as anything else. I think that this move derives in part from the implicit hope or belief that one metric would inform about both source-sink relations (getting the fruit ripe) and the cluster microclimate (ripening the fruit in the best environment possible); however, Y:PW does not accomplish either goal. Leading viticulturists acknowledge that the vine balance concept is poorly elaborated, and the arguments that HYLQ and vine balance are long-standing truths fared poorly upon investigation beyond the surface of the myth.

Similarly, no narrow period of ripening sensitivity to the environment has been discovered that could or should be called a "critical ripening period." There is also no reliable evidence of a narrow period of rapid changes in fruit composition that occur late in ripening, although that possibility has not been investigated thoroughly. What's more, assigning critical status to the end of the season misses the importance of the early season in determining several aspects of fruit composition, including important flavor components such as tannins, tartrate, and the veggy MIBP. A cohort of viticulturists relies on growing degree days (GDD) as though it were a fundamental parameter of plant-environment interaction like temperature or light intensity. However, it has long been understood outside of viticulture that GDD is an effective means of marking development when it works, and not effective when it doesn't. GDD appears to be an effective metric in grapevine development when the weather is relatively cool, but above only moderately high temperatures the metric has little or no direct relation to the rate of grapevine development or fruit ripening. Accordingly, a few key studies reveal the limited sensitivity of the rate of ripening to climate warming, compared to the well-documented sensitivity of the rate of vine development leading up to the onset of ripening.

The term "physiological maturity" (often employed in conjunction with the critical ripening period concept in winegrowing) is a novel if not corrupt application of a well-known and well-defined concept in agronomy and horticulture, where it refers to a physiologically defined stage *prior* to a final maturity—that is, prior to being ready to harvest (agronomic) or ready for use (horticultural).[2] Physiological maturity has no established basis in the physiology of the berry; its recent emergence to describe harvest decisions is a change in jargon that reflects a simple change in practice to later harvests (and higher Brix). In general, the rate of ripening slows as fruit reach maturity (as shown in fig. 2), which should render the harvest decision less (rather than more) critical for acquiring the optimum fruit composition. However, there are ongoing changes in grapes as they mature; rapid changes may occur that would include oxidation reactions as cells begin to die and the fruit eventually desiccate on the vine.[3] A physiological basis for physiological maturity could in principle emerge; however, that basis would necessarily be a physiological condition that correlates with a preference judgment, in terms of what the winemaker desires for that wine model.

The final, but perhaps most pervasive myth of winegrowing discussed in this book, is the terroir explanation for fine grapes and wine. I did my best to sort out what is going on with terroir—and was able to track an interesting progression in its use in winegrowing. The terroir notion seems to have evolved from the ancient and mistaken idea that plants arise from eating the right kind of soil, to a descriptor of an undesirable wine flavor believed to come literally from the soil, to a synonym for a broader environment, to an immutable aspect of the environment, and finally to its use in describing places and practices that give rise to wines that are cultural treasures. The widespread use of terroir in winegrowing today involves a 180-degree change in its connotation. Terroir, an old term used to localize wines, gained traction in conjunction with the well-publicized and well-documented increase in international competition in the wine market during the last decades of the twentieth century. Terroir and *goût de terroir* became a soil-based explanation for why some wine producers enjoy a reputation for the best, rather than the lowliest products. The insertion of terroir into the scientific literature also corresponded to the rise of terroir within the popular press. However, there is no consensus on terroir. Economists Rolf Mueller and Dan Sumner call terroir "a highly fashionable term which has as many shades and hues as a chameleon locked into a mirror box."[4] When terroir is used as a synonym for "good environment," a century-old truism results: the final fruit quality depends on the winegrape variety and its interactions with its environment. When used as a catchall term for every possible factor contributing to wine flavor (or a hypothetical distinctive flavor), the term conveys no useful information about the vineyard, and can be reduced to an acknowledgment that wines from different places are (often) just that: different.

Despite the lack of evidence in support of the concepts of terroir, HYLQ, BBB, vine balance, and critical ripening, these myths remain largely in place in the popular press as go-to explanations for wine quality. Myths die hard in winegrowing.

THE PERSISTENCE OF MYTHS IN WINEGROWING

The troubling evidence that runs counter to the myths of winegrowing does not appear in the popular press, where there is essentially no reference to the existing viticulture literature and exceedingly limited engagement with its authors. This isolation aids in sustaining the traditional explanations for fine grapes and wine. The reliance on wobbly concepts such as terroir, vine balance, and physiological maturity also contributes to their viability. One is much more likely to be correct when employing vague ideas than if one were to speak specifically,[5] but the myths are sustained by more than an elusive vagueness.

Truth in Most Myths

The myths explored in this work do contain important nuggets of truth. The most vital truth in the terroir explanation and the other myths of winegrowing originates in the history of poor vineyards, marginal climates, and limited knowledge of how to ameliorate problems in the vineyard. When the traditional thinking was taking hold, many vineyards were likely sick with diseases and poor nutrition that inhibited photosynthesis and fruit ripening. Add the challenges of end-of-season outbreaks of fungal diseases in the clusters, and we realize that the primary concern was simply obtaining a crop, rather than sorting out the nuanced flavor issues that are associated with quality today. With the proper environment (terroir), less crop (HYLQ or lower Y:PW), and a few more good weather days in the fall (critical ripening), fruit at that time were able to reach a more advanced ripeness than without those conditions. But for many of today's winegrowers, wine critics, and consumers, grape and wine quality is more than a gross concentration of solutes.

Beyond the role of crop load in the timing of ripening, less can be said with confidence about truth in the myths that is meaningful. There may be high yields beyond which fruit composition and wine quality suffer precipitously, but even there the fundamental issue may again be the capacity to ripen the crop rather than quality per se. There may be more color in red wines when smaller berries from a similar microclimate are selected, but these are merely speculations about HYLQ and BBB. Although it is clear that some amount of healthy leaf area (for photosynthesis) is required to ripen a crop and nourish the grapevine for next season, the vine balance concept has not been used to identify or refine that point. There are valid balances in grapevines, in root and shoot growth, for example, and thus, vine balance may find a more solid basis with respect to *fruiting,* as opposed to ripening. In this book, I argued that the original balance concept dealt with fruiting, but that old idea is also not being pursued by viticulturists today.

Fruit composition changes during ripening in ways that are important for wine flavor and aroma, and this makes clear that the harvest decision is important in determining aspects of wine flavor. By moving to heavier, higher alcohol wines obtained by later harvests, winemakers may have developed a feel for aspects of late ripening that will be borne out in future studies. Also, the presumed uncoupling of various ripening processes, in which warm conditions cause sugar accumulation to race ahead of other ripening processes, is not yet established but is being actively investigated.[6] The notion that a stressed vine makes the best wine probably holds in only a limited way for water deficits in red winegrapes, and to a lesser degree for vine nitrogen status. For both water and nitrogen, and indeed for environmental parameters in general, the evidence indicates that there is a target window of moderate deficit that secures the desired growth, productivity, and fruit composition, rather than a one-sided "more stress is better" principle.

Everybody Is an Expert

More important than any truths present in today's winegrowing myths is the commitment of wine experts to the traditional explanations for growing fine winegrapes, and their tacit acceptance by viticulturists. The received wisdom of winegrowing has been passed on as in the childhood game of "telephone," largely escaping the direct evidence-based evaluation that has been provided in this book. While fine wine is a gift to us from our predecessors, the traditional explanations for how quality originates in the vineyard are less so. Today's popular winegrowing myths are largely oversimplifications that require little or no insight into how the grapevine works. Because the major myths discussed in this book do not reflect the complexities of how grapevines operate, wine writers, winemakers, other wine professionals, and consumers outside of viticulture, can readily understand, accept, and repeat them.

Experienced wine appreciators have developed skill, resolution, flavor concepts, and memories, with a corresponding vocabulary to articulate their sense experience when tasting wine.[7] The talents developed in tasting acumen are not particularly useful in, for example, sorting out whether and how irrigation dilutes grapes or inhibits tannin synthesis. Wine journalists report their own taste experience and often connect those experiences to stories from the vineyards; the sources of those stories are mostly professionals in wine production. Similarly, winemakers invest heavily in learning to nuance their fruit toward desired outcomes in the bottle, but again talent in the handling of musts and wines does not translate into insight into grapevine water relations. While some in winegrowing "do it all," they seldom have the time and resources to test their own casual observations about grape development for validity as causal factors, although many are curious to know more.

In case I failed to make it clear earlier, in the conventional wisdom much of the logic regarding what makes a fine winegrape has been something along the lines of "This wine is great and has long been known to be great; therefore grapes should be grown as they are in that vineyard." That is tradition without thought, about both what the grapevine needs and what its environment offers. At times, reporters even denigrate deviations from traditional practices, but without recognizing the environmental contexts in which those practices developed. It is curious that the world of wine does not include a cadre of top writers from the pertinent disciplines. When one reads the work of wine writer Jamie Goode, who completed a PhD in biology, and therefore has some insight into how the grapevine operates and is trained in the analysis of empirical evidence, there is a markedly more noncommittal and measured reporting of the received wisdom.

Again, my sense is that members of the wine press report what their sources tell them, and in the end, the reporters and their sources have not done the work to critically evaluate the myths and legends they endorse. The philosopher Harry Frankfurt, author of the book *On Bullshit,* would call these orations just that. As Frankfurt notes, "BS is unavoidable whenever circumstances require someone to talk without knowing what he is talking about. Thus, the production of BS is stimulated whenever a person's obligations or opportunities to speak about some topic exceed his knowledge of the facts."[8] Suggesting there is bullshit in the world of wine is hardly novel. As mentioned in the previous chapter, Australian James Busby suspected the proverbial distinction between vineyards separated by a fence was "quackery" back in 1838. Economist James Quandt, who studies wine and wrote a paper on BS in wine evaluations, says, "I think the wine trade is intrinsically bullshit-prone and attracts bullshit artists."[9] When I told the winemaker at one of Napa Valley's leading midsized wineries that I was working on a book that dealt with bullshit in winegrowing, he responded with a chuckle and asked, "How are you going to know when to stop?"

When academics adopt what they hear, unconsciously skipping over the part where they test ideas or form their own opinions based on evidence, they have failed. I cite one clear example of my own such failure in the chapter discussing berry size. The Big Bad Berry myth was present in the collective winegrowing conscience with nary a published experiment to test it when I jumped in with both feet, interpreting my fruit color data according to the myth's predictions.[10] Later, I realized that I did not have enough information about the berry to make that interpretation, and began to make the appropriate measurements to evaluate the relative amounts of skin and flesh and their constituents.

Vested Interests in the Status Quo?

Is it possible that some wine experts don't care to know how the grapevine operates? The status quo works well for many in wine production, sales, and criticism. Wine is held in high regard by many, and the high end of the market is in the stratosphere. There is a parallel in the staying power of both the myths of winegrowing and the traditional wine hierarchy, and in investigating these myths, I've found that the perpetuation of the popular winegrowing myths may owe more to tacit acceptance that fits wine business concerns than to disinterested investigations in the vineyard.

The grandfathering of wine quality perceptions and positive associations with certain places and producers, limits to the production of most wines, and discounting of the importance of grapevine variety and winemaking practices (that can be repeated anywhere) all contribute to maintenance of the status quo. Attributing wine quality to local conditions is precisely the approach to take if one fears that wine of a similar quality could be produced elsewhere. Thus, the traditional explanations may simply be preferred in a world that does not concern itself with careful thinking about grapevines, but does invest heavily in rational thinking about wine business.

In this world of wine, the specifics of terroir (such as whether any specific soil or climate characteristic can be tied to a wine flavor, aka typicité) are a minor concern for an established producer when compared to the whole enchilada—the concept of identifying the wine with a place, or "terroir." Hence, the continuing mishmash of terroir definitions. Geographer and terroir scholar Warren Moran says that terroir begins when someone is selling wine,[11] and I concur. A farmer knows and deals with specifics of soil, climate, weather, varietal characteristics, canopy management, pest and disease pressure, and so on. After harvest, the crop is made into wine, and a label is attached—it is then that terroir appears as an extrinsic characteristic of the wine.[12]

Of course, wine producers are not unaware of this. A 1999 issue of *The Economist* featured a piece on terroir that included this story:

> In a recent lecture Michael Paul, the head of European operations for Southcorp [now part of Australia's Treasury Wine Estates], conjured up an imaginary region with a fragmented industry that sounded remarkably similar to many parts of Western Europe. Mr. Paul's fictitious region had managed to overcome its inherent disadvantages by virtue of a strategy dreamed up by a marketing genius. This involved popularizing the idea that wine was the product of the soil, the microclimate and the aspect of the vineyard rather than the wine maker. . . . The fictional marketing man's code name for the project was SCAM, an acronym of Soil+Climate+Aspect = Mystique, but in public he used the word terroir, which sounded earthy and natural, and had the additional advantage that few people really knew what it meant but they all felt they had to pretend they did.[13]

Whether Mr. Paul's hypothetical reflects a reality in terms of an intentionally misleading marketing plan or not, in the end, *terroir* is a term used mostly in wine marketing or in conjunction with the sale of grapes or wines as part of the story of a wine. In these contexts, *terroir* is used by people with more interest and expertise in wines than grapevines, and therefore a vague term will do. Paul does raise the notion of mystery, which may reflect a belief that there is a self-interest in not knowing more about how and where fine grapes and wines can be produced. The prospect of low yield, a sense of struggling vines, and the mystique of the winemaker's critical decision at harvest fit well into this scenario of exclusivity, whereas seeing the potential sites for fine winegrowing expand would be counterproductive.

When *terroir* is used in the context of viticulture, it is most effective as a synonym for *environment*, in which case using the term *environment* would be clearer and more accurate. Terroir introduces more confusion than clarity to the pursuit of fine winegrapes and is not needed in the study of grapes and wines. While possibly giving the appearance of knowledge and thus being effective for marketing (much like the emerging myth of biodynamics), no insight into grape development can be gleaned from the concept. Furthermore, terroir's claims to site-derived quality continue to be challenged by evidence that points to variety and winemaking-derived factors as key quality determinants.[14]

In addition to consumers, critics, and producers, academics enjoy wine and the wine culture as well. I'll admit, it is fun conducting experiments in famous vineyards (and in famous people's vineyards), and some nice wine occasionally passes hands. However, my own experience at a few viticulture and enology research meetings was that the atmosphere differed from other crop and plant science meetings in an uncomfortable way that smacks of Steven Pinker's description of the psychology of arts and psychology of religion, where the fields have "been muddied by scholars' attempts to exalt it while understanding it."[15] For some viticulture researchers, operating in the midst of fine wine and the wine culture may unconsciously guide experimental questions and interpretations toward conformation with the prevailing wind, and away from direct investigation of conventional wisdom.

LIFE WITHOUT TERROIR AND OTHER MYTHS

This book began with my desire to understand the received wisdom on winegrowing. At the time, I had no idea of how big a challenge I was undertaking. That being said, there are additional concepts beyond the myths covered here that are invoked as keys to winegrape quality, including vine age, fruit uniformity, and choice of rootstock. Although these concepts have not been thoroughly investigated, the wheels really start to come off with the increasing adoption of biodynamics. Biodynamics is a forebear of today's organic movement; however, it is based on occult ideas about water and nutrients that employ homeopathy, buried cow horn, and an astrological calendar. Biodynamics originated in a 1924 lecture series on agriculture given by Rudolf Steiner. Steiner was a polymath who initiated the Waldorf schools and the esoteric philosophy called anthroposophy, but he had no experience with agriculture.[16] In his few agriculture lectures, Steiner presented ersatz ideas about what makes plants grow. He recommended, for example, a particular way of stirring water to align the molecules before applying the water to the soil. According to his protégé Ehrenfried Pfeiffer, Steiner said that "plants themselves could never be diseased in a primary sense, 'since they are the products of a healthy etheric world.' They suffer rather from diseased conditions in their environment, especially in the soil; the causes of so-called plant diseases should be sought there."[17] With biodynamics, we are brought back not only to the soil and vitalism, but to mysticism as well. Biodynamics arose from the clairvoyance claimed by Rudolf Steiner—he believed that one could "just know" things, such as how Steiner *just knew* crop plants would benefit from a properly stirred slurry.

The recent growth of the biodynamic movement in winegrowing is indicative of thinking in the world of wine that tends to cling to mystery and traditional beliefs. It is not surprising that it is in winegrapes, not raisins or table grapes, that biodynamics has made the most inroads in viticulture.[18] If successful there, we may see coffee follow, as it has with terroir.

In short, viticulture could just as well drop terroir and the other myths of winegrowing. We clearly do not know as much about winegrowing as the conventional wisdom indicates, because predictions of the popular myths are so often not realized. At the same time, today things are better in the vineyard and in the bottle than we could have arrived at by blindly following the received wisdom of winegrowing. There is much fine wine today produced in more places than before and not because we have produced lower and lower yields, smaller and smaller berries, or planted on increasingly exceptional soil (or rock).

The linguist Noam Chomsky is widely cited for suggesting that our ignorance can be divided into problems and mysteries, in that "when we face a problem, we may not know its solution, but we have insight, increasing knowledge, and an inkling of what we are looking for. When we face a mystery, however, we can only stare in wonder."[19] Certainly, there's no mystery in seeking the highest Brix possible before harvesting. We are similarly faced with little mystery when it comes to evaluating the impact of environment and farming practices on grapes. To a viticulturist, these problems stand as interesting opportunities to contribute to the advancement of winegrowing. Winegrowing has advanced along with modern agriculture to farm grapes better.

This review of the major myths of winegrowing does not attempt to bring the reader up to date on the environmental biology of the grapevine, and the role of improved pest and disease control was underplayed in this book. But at the end of the day we have healthier grapevines and better grapes because of increased understanding of the temperature, light, water, and nutrient requirements of the grapevine. These healthier vines produced the long-term trend toward higher yields and better wines from riper, sounder grapes. A result is that even with greater yields, the concern for attaining the fruit maturity required for fine wine is diminished, if not eliminated in some regions, especially where climate warming has enhanced the length and warmth of the season. Further refinements in vineyard management that improve winegrapes have come from attention to managing the cluster microclimate and early-season vine-environment interactions (e.g., fruit exposure to light and vine water status).

Nevertheless, there is still much more that is unknown about optimal winegrowing conditions than is known, and it is helpful to acknowledge our limited understanding. The retelling of the popular myths has molded today's common understanding of winegrowing, even shaping the views of some scholars. The popularity of these myths contributes to the prevailing wind (illustrated in Hillel's "Path to New Knowledge," fig. 4) that impedes the pursuit of knowledge of the grapevine that would promote quality, efficiency, and innovation in the vineyard. The received wisdom restricts progress, in part by discouraging questions (which promoters claim have already been answered) and constraining interpretations and exposure of relevant evidence.

When approaching a problem as complicated as winegrowing with a "beginner's mind," more possibilities are available, leading to more options for sites, variety selection, and cultural practices. New clarity about how the grape ripens cannot be generated from an armchair,[20] and research investment in viticulture has been remarkably low when compared to other crops and to the value of the industry,[21] despite many studies that show return on investment in agricultural research is very high.[22] Incredibly, the role of yield in grape and wine quality is not understood (or even studied extensively) despite its fundamental roles in grape sales and wine supply. The available research falls short of what is needed to make informed decisions in the vineyard.

Although absent from the myths covered in this book, technology in winegrowing is traditionally dismissed by wine elites as leading to somehow unnatural wines.[23] I am a big fan of not fixing what isn't broken, but new is not bad by its very nature. In this instance, keeping a beginner's mind may lead to more and better grapes at less cost (another win for the consumer). Mechanized harvests, once the bane of winegrowing for its rough treatment of grapes, are now on track to surpass hand picking in terms of the quality of the fruit that reaches fermentation, as new optical sorters can evaluate each berry, whereas hand pickers evaluate clusters.[24] Furthermore, these sorters will help resolve the BBB issue at least for berries grown alike in a vineyard, because optical sorters of mechanically harvested fruit will facilitate large-scale sorting of berries for fermentations based on size.

Another technology that winegrowing has failed to take full advantage of is modern breeding. The genetic diversity in winegrapes is large, but only a few varieties are widely grown, leaving them open to wide-spread problems with pests and disease.[25] This is another hindrance owing to tradition—the adherence to and reverence of traditional varieties—as though the epitome of adapted varieties and wine flavor potential was reached prior to any understanding of genetics. One of the most important paths forward will be to correct the century-long failure to employ classical and modern genetics in the discovery of improved winegrapes that are better adapted for ripening to specific climates, and possibly with new sensory attributes in the fruit.[26]

The fact that vines are healthier today means that for many vine-yards, the choice of when to harvest has become more of a wine style decision than a dangerous dance with end-of-season weather. The improved ability to grow ripe grapes allows the current controversy over late-harvested, high-alcohol wines to be played out. A related anecdote of my own comes from a phone call that I received a couple of years ago from the winemaker at one of Napa Valley's most famous Cabernet Sauvignon producers near Oakville, California. After he introduced himself, I thought, "Oh no, here we go," because I had recently given a talk in which I challenged the HYLQ paradigm. However, he was calling to tell me that as the winemaker overseeing the production of other people's grapes, he was all for less yield, but after he purchased his own property and walked his own rows, he began to question the wisdom of dropping what appeared to be perfectly good clusters on the ground.

My recent experience tasting an earlier-harvested Cabernet Sauvignon wine took me back thirty years in a very pleasant way, to wines that I enjoyed as I began my career. (Maybe the fruit maturity we used to harvest at back then was just fine after all.) Higher yields delay sugar accumulation leading to longer hang times, and this should be investigated for its possible benefits in fruit composition—and in leading to more (and therefore more accessible) fine wine. Most wine lovers enjoy well-made wines, including all various formats—sparkling, port, rosé, fresh crisp or aromatic whites, finesse reds, and now that they are produced, big reds. New styles have and therefore can again arise, and each case of a new style is a divergence from tradition. Academics have helped pave the way to healthier vines and wines, in some cases via new technologies, but it is the winemakers and other wine professionals who set the paths of wine styles. The ongoing push to understand how winegrapes develop only adds to the proficiency with which producers can meet their stylistic goals.

Wine producers should not worry that wine will be less interesting (or in less demand) when we understand the grapevine better. Humans like alcoholic beverages, wine often enjoys the highest rank among them, and wine's appeal continues to grow. Wine won't lose its sacredness if we endeavor to better understand how to produce what we like, although new and specific knowledge about how environments impact the final product could put some individual producers at a disadvantage, compared to when such knowledge was unavailable.

Furthermore, there was nothing revealed in this work to prevent the honest marketing of wines based on their place of origin. There is much evidence that similar wines can be produced from different places, and that winemaking decisions play a powerful role in the sensory attributes of the finished product. A recent study of Bordeaux wines led to the conclusion that terroir played no role in distinguishing the wines;[27] however, there is no way to extrapolate from those wines to the rest of the world of wine. Many wine experts feel they can identify the source of wines, at least those wines from some regions/appellations. This could be true for reasons that involve the whole production system, including site, variety, vineyard management, and winemaking. One way of gaining insight into the relative role of the grape versus the winery is to look more closely at the grapes, rather than skipping over them to the wines, to see whether a characteristic of an environment consistently results in a characteristic in the fruit. Discerning more about the role of the site is an important exercise for those interested in resolving the relative roles of the environment, skill in the vineyard, and skill in the winery. However, a definitive answer will be a very long time in coming, in part because there is so much in the fruit to which we still cannot assign a role in the wine.

Adopting more of a beginner's mind may be helpful as a wine appreciator as well. Although there is evidence that our taste experience is somewhat dependent on the story we take with our wine,[28] less reliance on the extrinsic wine factors (including whatever myths or received wisdom the label or authority suggests) will open up new possibilities for discovering and tasting fine wine—perhaps from unexpected places. More attention to the taste of the wines and less to the pedigree of the label led to the rise of Robert Parker's consumer advocacy (from the early 1980s) that has successfully challenged an entrenched wine hierarchy.[29] Notice how the timing of yet another threat to the establishment (as with the 1976 Judgment of Paris tasting and the increase in international wine competition) corresponds with the rise of the terroir explanation. I think there is an ongoing power struggle over who determines the wine models of excellence that winemakers work toward, and some power has moved from the producer who tells the wine lover what is good to the (somewhat) independent voices of the wine press. With the development of social media, more power is moving toward consumers, and iconoclastic wine expert Tim Hanni is working to help them understand and follow their own preferences rather than his or that of another wine professional.[30]

With an open mind that doesn't rush to be an expert, a willingness to challenge the status quo, and the desire to investigate for the truth in the received knowledge of winegrowing, there is much more to learn about growing quality grapes. New knowledge will provide opportunities for innovation in the vineyard and perhaps for new sites to be discovered, where both traditional and new stylistic goals can be realized.

Notes

INTRODUCTION

1. McGovern 2003.

2. "Myth" is used here to refer to a traditional story that serves as an attempt to explain what would otherwise be mystery. As such, the explanation may be true, false, or somewhere in between.

3. The terpenes typically are described as having floral aromas, and sometimes as green or veggy.

4. Phenology is the study of weather or climatic regulation of biological development in time. In viticulture, budbreak, bloom, veraison, and fruit maturity are the main phenological stages.

5. Klebs 1910; Bopp 1996; Kramer 1956.

6. I use "variety" because that is the more commonly used term, although "cultivar" is usually the more accurate term. Within varieties, there are often several "clones"; these are plants that are derived from one variety and maintained by vegetative propagation, but that have some real or perceived differences in their characteristics.

7. Peynaud 1987, 222–25 is just one example of Peynaud's thoughts on this matter.

8. I use "models" here as Harry Paul did in *Science, Vine, and Wine in Modern France* (1996) to refer to the ideal to which the growing and making of a wine style aspires.

9. Who should decide is an important question, and it is the wine producer who has traditionally "educated" the wine consumer. As with other aesthetics, there is tension between the artisans and the critics, and it has increased as the wine critics have assumed more power in this relationship.

10. Hillel 1987.

11. Descartes 2001, 15.

12. Einstein 1934.

13. Cromer 1995. Physics professor Alan Cromer calls this egocentric thinking and argues that it interferes with our ability to learn nonintuitive aspects of the natural world.

14. Peynaud 1987, 223.

CHAPTER 1. LOW YIELD AND SMALL BERRIES DETERMINE
WINE QUALITY

1. Dresser 1992.

2. From a talk given at a Sonoma County, California, growers' meeting at which we were both speakers.

3. Shanken and Matthews 2010.

4. Hooke 1991.

5. See the description of the Zinfandel Heritage Vineyard Project, established at the Oakville Station of the Dept. of Viticulture & Enology, Univ. of California Davis: http://www.zinfandel.org/default.asp?n1=18&n2=787&member=

6. Keller et al. 2005

7. Jackson and Lombard 1993.

8. For example, see Virgil, *Georgics* 2.112–13 (trans. Kline). In this case, the translation is "lastly Bacchus's vine loves open hills."

9. Smart 1992, 85. The Bacchus quote appears again on p. 190 of the companion volume, *Resources,* where it is interpreted as the conceptual fruit of centuries of trial-and-error site evaluations.

10. Robinson 2002a, in her review of *The Far Side of Eden* by James Conaway.

11. Robinson 1999, 786–87. The section on yield carries the initials of Richard Smart and Jancis Robinson.

12. Virgil, *Georgics* 2.109–13 (trans. Kline).

13. Virgil, *Georgics* 2.184–94 (trans. Kline).

14. Earlier examples are Jean Baptiste Francois Rozier (1777), in *Introduction aux observations sur la physique, sur l'histoire naturelle et sur les arts, avec des planches en taille-douce, dediees a monseigneur le comte d'Artois,* 457; and Jean Louis Alléon Dulac (1765) in *Mémoires pour servir à l'histoire naturelle des provinces de Lyonnois, Forez, et Beaujolois,* 189. In another myth, Chaptal is often given credit for inventing the "Chaptalization" process—adding sugar to the juice of winegrapes to feed the fermentation when fruit failed to ripen adequately. Although his name is used, this practice did not originate with him as he himself explicitly said; see Gough 1998; also Paul 1996, 123–54.

15. Chaptal 1801.

16. From Redding 1833 and Husenbeth 1834 to Coyle 1982 and New York State Agricultural Experiment Station, and New York Wine Grape Foundation 1990.

17. Wilkinson 1978, 50–60.

18. Seneca, *Letters,* p. 149 (trans. Campbell).

19. Cato the Censor, *De agricultura,* chap. 1 (trans. Dalby); see also Unwin 1996, 102–4.

20. For example, a recent survey (Cai and Noel 2013) placed vineyard land value at \$56,000–\$220,000/acre, and the second highest crop land was in strawberries at \$29,000–\$51,000/acre.

21. See Pliny, *Natural History,* bk. 14, "The Natural History of Fruit Trees."

22. Columella, *De re rustica* 4.3 (trans. Ash).

23. Columella, *De re rustica* 2.15.1; 2.2.25; 5.5 (trans. Ash).

24. Pliny, *Natural History* 12.4 (trans. Bostock and Riley).

25. Columella, *De re rustica* 2.15 (trans. Ash).

26. Pliny, *Natural History* 14.1 (trans. Bostock and Riley).

27. Pitte and DeBevoise 2008; and on a few commercial wine websites.

28. Berlow 1982.

29. Ibid.

30. Ibid.

31. Unwin 1996.

32. Quoted in Haeger 2004, 17.

33. It has been argued that scarcity of manure limited agricultural production in general in the fourteenth century; Dyer 1989, 41. Moreover, it is argued, and supported in various ways, that peasants probably favored their own gardens over the fields of the estate in their use and application of manure. See, e.g., Astill and Langdon 1997, 20; and Dodds and Britnell 2008, 82.

34. Loubère 1978, 126.

35. Olivier Serres's *Théâtre d'agriculture,* cited by Pitte and DeBevoise 2008, 14.

36. Bell 1980.

37. Loubère 1978, 125.

38. Weigend 1954.

39. Nemani et al. 2001 reported that Napa yields have increased dramatically, but I present Napa County yields in fig. 6B that conflict with that assertion.

40. Unwin 1996, 360.

41. Toussaint-Samat and Bell 1992, 288.

42. In the United States, boundaries delimit American Viticultural Areas instead of appellations.

43. Gangjee 2012, chap. 3, "The Appellation of Origin in France."

44. Oczkowski 2006.

45. Coleman 2008, 45–48.

46. Smith 2004.

47. Predicting yield is notoriously difficult in vineyards. In France and other parts of Europe, yield is back-calculated from the volume of must or wine.

48. Kevany 2008; Stevenson 2007.

49. There are other types of contracts, including a price per acre and a price determined by wine "bottle price," which are intended to free the grower from the motivation to produce high yield.

50. A few wines are now at suggested retail of more than \$100,000—including an Australian Penfolds, for example; see Divirgilio 2012.

51. Robert Parker is widely acknowledged to be the most influential wine critic in the world today. Parker received worldwide attention when he called

the 1982 vintage in Bordeaux superb, contrary to the opinions of many other critics. Parker began as a consumer advocate (à la Ralph Nader) and has been critical of most wine writers for their close relationship with wine producers.

52. See Parker 1997, 27; and later books by Parker.

53. Parker 1997, 21; the same quote is repeated in books by Parker from 1997 to 2008.

54. Halliday and Johnson 2007, 54.

55. It is remarkable to me that Halliday and Johnson imply wines were better in 1870, but we cannot know one way or the other. As both Halliday and Johnson and Parker refer to the better "staying power" of earlier wines, the sense that Rieslings of the 1870 were better may come from the wine experts' belief that those earlier wines had longer lives. The arc of a wine's flavor over time and whether the length of that arc is quality are interesting issues that fall outside of the vineyard focus of this book. However, I will note that wine producers have purposely moved to producing wines that are ready to drink sooner.

56. Wine scores from Sotheby's were presented by Nemani et al. 2001, who reported that those and other vintage ratings showed increasing scores from approximately 1970 to 1996, and independent analysis, which included *Wine Spectator, Wine Enthusiast,* and *Wine Advocate* data, indicated the same trend. We (Mark Matthews, David Block, and Charles Reisman) tested for and found no significant variation among the three critics ($p = 0.17$, NS) (unpublished).

57. In addition to the vineyard yields, there is the very significant question of how reliable the wine scores are. Some analyses of wine critics indicate they are consistent among themselves (e.g., Nemani et al. 2001), and others show that not to be the case (Quandt 2007a). There may be more agreement when the wine press is evaluating vintages rather than specific wines.

58. Halliday and Johnson 2007, 56.

59. Berry size in Riesling tended to increase when clusters were thinned (Preszler et al. 2013), or when berries were thinned by applying a hormone (Weyand and Schultz 2006b), although in neither case was the change large. Robertson et al. (2009) showed a linearly increasing acidity with increasing berry size in Pinot noir (see fig. 3).

60. Reynolds et al. 1994b.

61. MacNeil 2001, 875.

62. Ibid., 29.

63. Halliday and Johnson 2007, 50–52. When a scientist invokes a "direct relationship," a linear one like curve 1 is usually implied, and a causal relationship is suspected.

64. MacNeil 2001, 281.

65. Ibid., 822.

66. Ibid., 29.

67. Jackson and Lombard 1993. The original research is cited in their review article.

68. Zamboni 1996.

69. Peterlunger et al. 2002.

70. Reynolds et al. 1996a.

71. Heazlewood et al. 2006.

72. Halliday 2006, 78.
73. Freeman et al. 1980.
74. Freeman 1983.
75. Bindon et al. 2008.
76. Lawther et al. 2010, 27–28.
77. Brook et al. 2011, 33; Osborne 2004.
78. Olmo 1979; Winkler et al. 1974, 195.
79. Hanson 1995, 167–68.
80. Shaulis and Kimball 1955; Wiebe and Bradt 1973; Archer and Strauss 1991; Hedberg and Raison 1982.
81. Archer and Strauss 1991 noted Ravaz's early work: "Ravaz (1908) has proven that different vine spacings induced different rates of shoot growth and that the longer growth period of wider spaced vines exerted a negative effect on grape quality." However, no proof is evident. In fact, Ravaz's data do not support that notion, even by his reckoning (see Ravaz 1908, 285–86), although Ravaz did seem to assume that narrow spacing is better.
82. Pastor et al. 2007; Beuerlein 2001.
83. Winkler 1959, 1969. In studying the same vines, Winkler reported lower yield in the most dense spacing compared to eight other spacings for Riesling and Cabernet Sauvignon in Napa Valley in 1959, but not when evaluated again 1969. The results of that study were unusual in that yields were statistically similar among a wide range of spacings, suggesting that the vines were managed to obtain similar yields.
84. Simon 2003, 85.
85. Edwards et al. 2009, 28.
86. Archer and Strauss 1990.
87. Williams and Arnold 1999. There were no significant differences in berry size, but a trend to smaller berries with closer spacing.
88. Turkington et al. 1980.
89. No effect: Archer and Strauss 1991; Hunter 1998; Reynolds and Wardle 1994; Hedberg and Raison 1982; and Intrieri 1987. Riesling: Reynolds et al. 2004b.
90. Barbera: Bernizzoni et al. 2009; Chancellor: Reynolds et al. 1995; Cabernet Sauvignon: Kliewer et al. 2000. Similar results (little or no response) in studies with Merlot: Murisier and Ferretti 1996; Murisier et al. 2005. In Riesling, terpene concentrations were unaffected by spacing in three out of four years, and the results were inconsistent among types of terpenes in the other year, one type being higher in close spacing and the other type being lower in close spacing. Reynolds et al. 1996b.
91. Reynolds et al. 2004c. Reynolds and Heuvel 2009.
92. Freese 1999, and comments at the 1999 ASEV symposium, p. 23.
93. Hodson 2011; Morton 2011; Wolf 2011; Hill 2011. The divergent conclusions or recommendations in these studies are representative of the lack of consensus on spacing and quality, and on how to arrive at an optimum vineyard density.
94. Ted Lemon, quoted in Brook et al. 2011, 33.
95. Reynolds et al. 2004c.
96. Shaulis 1982; Smart 1985.

97. The fundamentals of spacing considerations and production implications are laid out nicely in Grant 2000.

98. Ravaz 1904a.

99. Weaver et al. 1957, 1961.

100. Freeman 1983.

101. Freeman and Kliewer 1983.

102. Kliewer 1991.

103. Heazlewood et al. 2006; see also Hunter and De la Harpe 1987.

104. Gray et al. 1997: wine value index; Holzapfel et al. 1999; and Clingeleffer et al. 2001. The range in yields was substantial: for example, 8–48 t/ha for Shiraz and 8–51 t/ha for Cabernet Sauvignon (Clingeleffer et al. 2001). Bramley and Hamilton (2004) concluded that there was no evidence that the lowest-yielding vines produced the best-quality fruit at Coonawarra and pointed to the irrationality of applying a constant yield limit in every season.

105. Keller 2005; Keller et al. 2008, 2010.

106. Freeman and Kliewer 1983; Chapman et al. 2004a, which includes an important example with MIBP and veggy flavors.

107. Wine astringency example: Chapman 2004; example of larger responses: Intrigliolo and Castel 2011.

108. Sauvignon blanc: Naor et al. 2002.

109. From Kliewer and Weaver 1971 to Nuzzo and Matthews 2006, and many others in between.

110. Wolf et al. 2003.

111. Cordner and Ough 1978; Hunter and De La Harpe 1987; Naor et al. 2002.

112. Sinton et al. 1978. Brix, pH, and nitrogen of the fruit were progressively delayed by increasing yield.

113. Bindon et al. 2008; Bowen et al. 2011; Bravdo et al. 1985a (for Cabernet Sauvignon); Cordner et al. 1978; Ewart et al. 1985; Freeman 1983; Freeman et al. 1980; Heazlewood et al. 2006; Hill 2011; Hodson 2011; Morton 2011; Peterlunger et al. 2002; Reynolds et al. 2004b, 2004c; Wolf 2011; Zamboni et al. 1996.

114. The term "vintage year" seems to be fading from use, perhaps as global warming and prudent viticulture practices reduce the need for supplementing sugar, and international competition becomes increasingly important compared to the weather. See note 14 in this chapter, on Chaptalization.

115. Hirsch 2005.

116. Parker 2008, 32.

117. The amount of solute can be expressed as weight or as the number of molecules in moles; the amount of solvent is usually expressed in volume. Thus, wine concentration can be expressed as mg/liter (L) or as molarity, moles of malate/L of wine. In biology and winegrowing, concentration in berries can be expressed similarly, but it usually refers to the ratio of amount of dissolved solute/fresh weight of the tissue or organ, where the fresh weight is used as the denominator instead of the volume of water per se. Like most living organisms, plant tissues including berries are mostly water—85 percent of berry fresh weight is water, although when using the fresh weight basis other factors such as the

amount of cell wall come into play. In wine, the alcohol can be considered solute as well as solvent.

118. Zorrilla-Fontanesi 2011.

119. Johann Six, UC Davis professor of agricultural ecology, personal communication.

120. Phillips 2003.

121. Clarke 2006, 35.

122. The extent of solute export from fruit is not known.

123. Alonso-Cantabrana et al. 2007: "Carpels and leaves are evolutionarily related organs, as the former are thought to be modified leaves."

124. There is a well-known phenomenon of lower protein concentration in higher-yielding alfalfa, but this has been shown to arise from an increased amount of stems relative to high protein leaves, rather than lower leaf protein concentrations. Orloff and Putnam 2007, 17.

125. Loughrin and Kasperbauer 2001.

126. For example, fig. 2.34 in Mengel and Kirby 2001, 103.

127. Kliewer 1971; Spayd et al. 1993.

128. Greenspan et al. 1994, 1996.

129. Murisier et al. 2005.

130. Rizzini et al. 2009; and other Pietro Tonutti works on gene expression in desiccating peach and grape.

131. MacNeil 2001, 18, 841.

132. Johnson 1989, 64–67; Petit 1985, 24.

133. MacNeil 2001, 18.

134. Coombe and Monk 1979.

135. There are two systems of conduits that transport water and solutes. One, called xylem, mostly operates delivering water and mineral nutrients from the roots to the shoots. The other, called phloem, mostly operates delivering water, sugars, and amino acids from photosynthesizing leaves to other organs including grape berries.

136. R. E. White 2009, 38.

137. These data were generated in 2005; a more recent analysis produced similar results.

138. Amerine and Roessler 1952. This clear and concise summary is still helpful today for those not trained in sensory science. Amerine and Roessler also initiated the food-based terminology used in wine descriptions.

139. Descriptive analysis techniques involve selection of panel members, term generation based on the wines, concept formation by training with standards, and wine sensory evaluations.

140. Bravdo et al. 1984, 1985b.

141. Data from Matthews and Anderson 1989 and Matthews et al. 1990.

142. Smart and Coombe 1983.

143. Chapman et al. 2004a.

144. The results of cluster thinning here were similar to several other recent studies that found little or no effect on fruit composition: Fanzone et al. 2011; Keller et al. 2008; Keller et al. 2005; Reynolds et al. 2007.

145. Chapman et al. 2005.

146. Reynolds et al. 1996a.

147. Chapman et al. 2004b.

148. In Sauvignon blanc, there are producers and consumers who go for and against significant vegetal character in their fruit and wines.

149. http://www.hesscollection.com/one-pager/napa-valley-chardonnay-2013/; http://www.hesscollection.com/wine/mount-veeder-cabernet-sauvignon/; http://www.professorshouse.com/food-beverage/wine/articles/colome-vineyards-argentina-a-hess-collection-winery/; from Hess Collection descriptions of their Cabernet Sauvignon Mount Veeder 2007, 2008, and Hess Collection Napa Chardonnays. The Hess Collection and Ridge produce consistently highly rated wines.

150. Coombe et al. 1987; Possner and Kliewer 1985; and Storey 1987, for potassium. There are important mutants that make color in the flesh, called *tintiere* varieties.

151. Loudon 1826, 71.

152. E.g., "small berries with an austere taste" are preferred for wine, in Lieber et al. 1832, 572.

153. Munson 1909, 126.

154. *Vitis vinifera* subspecies *vinifera* includes thousands of cultivars domesticated from the wild progenitor *Vitis vinifera* subspecies *sylvestris*. The wild subspecies *sylvestris*, widespread in Eurasia, has separate female and male flowers occurring in roughly the same proportion and small, acidic berries in comparison to today's cultivated grapes. Grape species native to North America, *Vitis labrusca, Vitis rotundifolia, Vitis aestivalis,* have given rise to winegrape varieties.

155. McGovern 2003, 33.

156. Aradhya et al. 2003 (describing the work of Negrul: Negrul 1938).

157. Winkler and Amerine 1943, 5.

158. Singleton 1972, 106.

159. Singleton 1966. Also, according to Vern Singleton in Singleton 1972, "On the other hand some muscat grapes, Semillon, and other varieties capable of making high-quality white wines have large berries . . . . A brief examination of several ampelographies gave an average berry diameter of 12.2 mm for 16 red and 12.8 mm for 15 white varieties noted for wine production and grown in California. While the two sets of values overlapped and were so variable as to preclude statistical significance without much more data, the findings agree with the [BBB] postulate." Singleton suggested that this accounts for a historical but unrecognized tendency to select toward small berry size for red wine varieties but not for white wine varieties."

160. Sugars are not higher in smaller-berried varieties: Poni et al. 2009; Weaver 1973; sugars are higher in smaller-berried varieties: Romero-Cascales et al. 2005.

161. Shellie 2007: 23-variety trial in Idaho; 13-variety trial in Oklahoma, data courtesy of Eric Stafne.

162. Reynolds et al. 2004a.

163. Ortega-Regules et al. 2006; Gris et al. 2010.

164. Watson et al. 1988.

165. Anderson et al. 2008.

166. E.g., Muller-Thurgau 1898; Scienza et al. 1978; Cawthon and Morris 1982.

167. Glynn 2003. A similar analysis for a Chardonnay cluster was reported by Trought and Tannock 1996. The same shotgun appearance to plots of Brix vs. berry size are present in data for Cabernet Sauvignon and for Cabernet franc taken from the variety trial in fig. 12B.

168. Kasimatis et al. 1977.

169. Roby et al. 2004.

170. Walker et al. 2005.

171. Coombe et al. 1987. The sugars in the ripe berry are almost exclusively equal parts glucose and fructose.

172. Roby and Matthews 2004; and Suklje et al. 2012.

173. Walker et al. 2005: concentrations of acids were highest or lowest in midsized berries, depending on the genotype of the roots; i.e., highest acids were not in the smallest berries. Although water deficits often reduce berry size and malate concentration, when water deficits cause berry size differences in Shiraz, the highest titratable acidity was in the smallest berries (Bindon et al. 2008).

174. Other reports with the assumption that the concentration of skin solutes in the must should follow 3/r: Gladstones 1992; Hardie et al. 1996; Matthews and Anderson 1988; Singleton 1972; Singleton and Esau 1969.

175. Bravdo and Naor 1996.

176. Kennedy 2002.

177. Hunter 1998.

178. Matthews and Anderson 1989; Matthews et al. 1990.

179. Roby and Matthews 2004.

180. Poni et al. 2009; Walker et al. 2005; Barbagallo et al. 2011; Storey 1987; shows no change in skin: flesh in berries of variety Tarrango up to 3 gm, except in very small berries under 0.5 gm.

181. Roby and Matthews 2004; Bindon and Kennedy 2011; Bucchetti et al. 2011.

182. Poni et al. 2009.

183. Bucchetti et al. 2011.

184. Roby et al. 2004; Walker et al. 2005; and Barbagallo et al. 2011; Nii and Coombe 1983.

185. Cortell 2006, 134.

186. Brook et al. 2011, 33; also http://www.hartfordwines.com/farming; http://www.kapcsandywines.com/about/management; http://www.opusonewinery.com/The-Wine/Vineyards

187. Archer and Strauss 1991; although mean berry size in the tightest spacing was 0.05 grams less than in the wider spacings.

188. Hunter 1998; berry size calculated from cluster weight and berry set data in table 2.

189. Reynolds et al. 1995; Kliewer et al. 2000; Murisier and Zufferey 2003.

190. Nardozza et al. 2010.

191. Keller et al. 2008; see fig. 4.

192. E.g., Turkington et al. 1980; and for pruning: Freeman et al. 1980; for cluster thinning: Winkler 1931; Dami et al. 2006.

193. Holt et al. 2008. Overall mechanical pruning caused smaller berries whose wines had significantly *lower* wine scores, see table 3.

194. Holt et al. 2008.

195. Roby et al. 2004.

196. Roby 2001.

197. Matthews and Nuzzo 2007.

198. Clingeleffer et al. 2002.

199. Walker et al. 2005.

200. Kasimatis et al. 1985; Johnstone et al. 1995.

201. Roby et al. 2004.

202. Dokoozlian 1990.

203. E.g., Kliewer and Antcliff 1970; Kliewer and Lider 1970; Hummell and Ferree 1998.

204. Bergqvist et al. 2001; Keller and Hrazdina 1998.

205. Walker et al. 2005.

206. Kliewer and Schultz 1973.

207. Lynn and Jensen 1966.

208. Kliewer and Weaver 1971.

209. Kliewer and Antcliff 1970.

210. Suklje et al. 2012; see, in Suklje et al., fig. 4, Controls sampled 1/3/11.

211. Holt et al. 2008 briefly reviewed several studies involving berry size in winegrapes and concluded that differing conditions influence berry weight and berry composition in different ways.

212. Roby et al. 2004 also showed that the water deficits increased the concentration of color and tannins at any given berry size, revealing effects of water deficits on fruit composition independent of berry size.

213. Di Profio et al. 2011. While high yield delays sugar accumulation in most cases, the literature is not consistent on the effect of cluster thinning on fruit acidity, which decreases during ripening. Fruit acidity is often considered marginally too high in cool climates and hazardously low in warm regions (where acid is sometimes added to the must). Di Profio (and Andy Reynolds) briefly reviewed the literature with respect to cluster thinning and fruit acidity, concluding that the most common response was little or no effect. Similarly, high pH in juice, musts, and wines is a common problem in warmer winegrowing regions. Conventional wisdom is that high pH develops in concert with high concentrations of potassium in fruit, which accumulates in the skin like color and tannins. It is interesting to note, however, that the conventional wisdom has not been to grow more fruit to reduce potassium concentration, and thereby juice and wine pH. Therefore, although I am suggesting that the primary consequence of reducing yield is advanced ripening, there is evidence that all the components of ripening are not affected in exactly the same manner.

CHAPTER 2. VINE BALANCE IS THE KEY TO FINE WINEGRAPES

1. Pool 2004; Dry et al. 2004; Howell 2001; Kliewer and Dokoozlian 2005.

2. Kramer 1983, 181, showed that, as in fig. 16A, nutrient deficiency favors root growth.

3. Petrie et al. 2000 show the linear relation between above- and below-ground growth in grapevines with various crop and leaf removal treatments. Kliewer and Fuller 1973 report that when leaves are removed, root growth is reduced. They also show a linear relationship between root and shoot growth in grapevines. The root pruning phenomenon described schematically in fig. 20B was observed in grapevines by Smart et al. 2006.

4. The SupAgro school is located on Ave du Professeur Louis Ravas in Montpellier, France.

5. Paul 1996, 28.

6. Viala and Ravaz 1888; the organism is *Guignardia bidwellii* (Ellis).

7. Howell 2001; Dry et al. 2004.

8. Maccarrone et al. 1996 and Scienza et al. 1996 cite a 1909 Ravaz publication on cultural practices, *Influence des opérations culturales sur la végétation et la production de la vigne.* The same work was apparently published under different covers in 1908 and 1909. It is possible that the Italian authors had also read the French text *Éléments de physiologie de la vigne et de viticulture générale* (Champagnol 1984), in which François Champagnol proposed target Y:PW values for a few winegrape varieties that he apparently extracted from Ravaz's work. Tassie and Freeman 1992, 73, cited Champagnol on this account.

9. Battistutta et al. 1997, 2000; Colugnati et al. 1997, 2000; and Stefanini et al. 2000.

10. Key words are words that are not in the title of published papers but can be used to help researchers locate relevant studies. Miller and Howell did not mention Ravaz in their work published in 1998 (Miller and Howell 1998), yet in a publication three years later, Howell included the Ravaz Index as a key word (Howell 2001).

11. Ravaz 1906; this publication reports primarily on the effects of soil color, vine spacing, and timing of pruning on must sugar, acid, and F/V (Ravaz used F/V [fruit/vegetation] for Y/PW). Ravaz 1911 reports the results of leaf-thinning experiments on must sugar and acid, in some cases with yield, and in a few cases with Y:PW data.

12. Vasconcelos and Castagnoli 2000 cite Ravaz 1903; that paper contains only some leaf mineral nutrition data and two values for must sugar: high for healthy vines and very low for vines with brunissure.

13. Ravaz 1902, Table I, pp. 482–83. He begins, "The cause of Vine Browning has remained, until now, unknown."

14. Ravaz 1904a, 86.

15. Ibid., 181. Ravaz 1904b also reported it was a malady of young vines. Weaver and Pool 1968 reported foliage browning in high-cropped mature vines, similar to Ravaz's report for young vines. Leaf browning probably has several causes.

16. Ravaz 1906.

17. Ibid. Tournier 1907, reporting on Ravaz, overproduction, and Pierce's disease, repeated this claim: "The California vine disease therefore is not properly speaking a disease at all and it is useless to say the least to make of it a scarecrow for viticulturists and to organize against it such a vigilant guard in all parts of Europe."

18. A. F. Woods (1899) said in an article in *Science* that he could reproduce what Viala saw in any plant, and did not need a causal organism or a sick plant to do it. According to Neuhauser et al. 2009, Massee showed this earlier and independently in 1893.

19. Goheen and Cook 1959.

20. Evidence pointing to diseased vineyards at that time (late 1800s to early 1900s) includes the well-known painting "The Red Vineyards near Arles" by Vincent Van Gogh. Supposedly the only piece sold by the artist while he was alive, the 1888 work depicts laborers in a vineyard containing only red and brown leaves.

21. Ravaz and Verge 1925; Ravaz et al. 1933; Zacharewicz 1934; Bonnet 1937; Maume and Dulac 1945; and Wilhelm 1950. In 1925, Ravaz distinguished red- from brown-leaf "disease," but he showed that red leaf, too, was corrected by potassium applications, just like brown-leaf.

22. As revealed in Ngrams of brunissure from Google Ngram viewer of digitized books.

23. For example, Bavaresco et al. 2010.

24. Surprising only in the sense of how his work has been portrayed, not in the sense that his work was directed at vine health rather than wine.

25. Ravaz 1935.

26. Husmann 1888, 102; George Husmann was on the viticulture faculty of University of Missouri and later moved to California, where he worked as an independent researcher.

27. "Vine balance" appears to be primarily a New World preoccupation. Although Ravaz's F/V metric was resurrected in a couple of French viticulture texts (Branas et al. 1946; Champagnol 1984), searches for *équilibre de la vigne* and related phrases in JSTOR, CAB abstracts, and Google France Scholar and Books all produce no hits. Searches for *l'équilibre vigne* in Google France Scholar and Books produced no hits. Searches for "Ravaz + vigne + F/V" also produced no hits. Searches on Google France produced many hits, all as *l'équilibre vigne-sol,* and mostly commercial wine sites such as www.domaine-rouaud.com, including biodynamic sites, "artmajor," lespassionesduvin, etc. On French Scholar or Books *l'équilibre vigne-sol* = no hits. Champagnol highlighted the target F/V values that he apparently extracted from Ravaz's limited data.

28. Ravaz 1904a, 112.

29. Dry et al. 2004.

30. Ross and Smith 1912, 108.

31. There is no evidence of chickens being at risk of dying from overlaying; U.S. hens produce more than two times more eggs/year than in 1910 and up to 300 eggs/year, but a two-egg day remains rare. However, the relative impact of improved varieties and improved farming practices on winegrape yields is not known.

32. Reference to vine exhaustion appears in several annual reports of the California Agricultural Experiment Station that were published from 1884 to 1914. For example, see Bioletti et al. 1886, 158.

33. Bioletti et al. 1886, 158; Bioletti 1913, 22.

34. Ravaz 1906.

35. Watson and Riha 2011.

36. Ehrlen and Lehtila 2002.

37. Munné-Bosch 2008.

38. A 1903 Michigan horticulture report attributed premature exhaustion in plums to overbearing caused by grafting onto a Japanese rootstock.

39. Old grapevines in Slovenia: http://maribor-pohorje.si/the-old-vine—the-oldest-vine-in-the-world.aspx#You%20don%6ot%20believe%20it; other oldest vines include the "Versoaln" outside of Merano, Italy: http://www.trauttmansdorff.it/en/news/archiv/2013/august.html; "The Great Vine," a nearly 250-year-old Black Hamburg vine located at Hampton Court in London: http://www.hrp.org.uk/HamptonCourtPalace/stories/palacehighlights/TheGreatVine; and "The Mother Vine," a variously claimed 300- or 400-year-old Muscadine (different species) grapevine on Roanoke Island, North Carolina: http://www.northcarolinahistory.org/encyclopedia/122/entry.

40. Clingeleffer 1984.

41. According to a Google Ngram of GoogleBooks.

42. Winkler et al. 1974, 300.

43. Johnson 2004, 122.

44. "Fruit trees are pruned to regulate growth, increase yield, improve fruit size and quality and reduce production costs. Pruning also shapes trees for convenience of culture and repairs damage." Herrera 2004.

45. Bond 2000.

46. Bioletti 1897, 4; Bioletti also referred to a "well-balanced" vine (12), but the meaning is not clear from the context. Bailey (1902, 860; 1916, 12, 90, and 109) refers to balance in vines with respect to relative root and shoot growth. Evidently, there was no "balance" in viticulture earlier; e.g., not in Mohr 1867.

47. Bioletti 1897, 2–4, and modified later in Bioletti 1913. Bioletti's "Vine Pruning" appeared in several volumes, as did some of Ravaz's works, suggesting that in contrast to today it was common practice at the time to publish the same work in multiple places.

48. Bioletti 1897, 25.

49. Winkler 1934 (revised 1945), 9; and in the 1962 text *General Viticulture*, 246.

50. Ravaz was aware of Winkler and Bioletti's work, and wrote in praise of it. Bioletti had many Ravaz papers in his files, so Winkler, who worked alongside Bioletti, must have in turn been aware of Ravaz's work. Yet while Winkler mentions other famous French viticulturists Guyot and Foex, he never refers to Ravaz or his Y:PW ratio—perhaps because he saw no development of a balance concept in Ravaz's work.

51. Partridge 1925, 37. Newton Partridge was a professor of horticulture at Michigan State University.

52. Thomas and Barnard 1937, as cited in Winkler et al. 1974 (*General Viticulture*, 2nd rev. ed.).

53. Hake et al. 1996, 101.

54. Williams et al. 2010a and b; shoot-length data taken from one paper, and yield from the other paper.

55. Boss et al. 2003; Boss and Thomas 2002.

56. As late as 1957 (Weaver et al. 1957), Winkler interpreted data as evidence that "two years of overcropping weakened the vine so greatly that very little crop was produced the following year." But a few years later, Weaver et al. 1961 concluded from similar experiments with high crop loads that "3 years of over cropping . . . failed to markedly weaken the vines."

57. Pruning to high number of buds: Chapman et al. 2004a; irrigation: Matthews and Anderson 1989.

58. Keller et al. 2004: four out of five years.

59. Budbreak over 100 percent can occur because the basis for the 100 percent is the number of buds that were left on the vine at pruning and expected to grow into shoots, and there are always large numbers of buds on vines that are not accounted for in pruning. These buds can always remain dormant, or grow when conditions promote their release from dormancy.

60. Clingeleffer and Krake 1992; Weyand and Schultz 2006a.

61. Ravaz 1906, 37.

62. Phillips 2014, 529. The Green Revolution was a period of dramatic increases in yield of wheat and rice in the 1960s based on new varieties and increased inputs. Norman Borlaug won the Nobel Peace Prize in 1970 for his contributions to increased food supplies.

63. Primack and Hall 1990, 638.

64. Obeso 2002.

65. Knops et al. 2007.

66. No viticulture or popular winegrowing publication makes reference to the cost of reproduction literature.

67. Water deficits: Matthews et al. 1987; Williams et al. 2010a and b; Nitrogen: Kliewer and Cook 1974, but not in Keller et al. 1998.

68. There is an impenetrable chapter on equilibrium (balance) in Branas et al. 1946.

69. However, in the 1920s, Partridge introduced what came to be known as "balanced pruning," in which the "correct" number of buds left on a vine at pruning time is determined by the pruning weight; e.g., leaving ten buds per pound of prunings.

70. Bravdo et al. 1984. Note that yields were indeed very high, 29.3–34.3 tons per hectare.

71. Bravdo et al. 1984.

72. For example, Santesteban et al. 2010, 51: "Several indices have been classically used to estimate vine balance; mainly pruning wood to yield and leaf area to yield ratios. Balanced vines are usually said to have Ravaz Index values between 5 and 10"; A. Lakso, as quoted in Chien 2012: "5–10 lb of fruit per lb of pruning weight is an optimum range." The 5–10 rule also appears in Extension Service publications; for example, Hamman and Dami 1999; Vance and Skinkis 2013; and in the online pruning guide from Washington State University, *Pruning Basics*, http://wine.wsu.edu/research-extension/2005/02/pruning-basics/.

73. Hepner et al. 1985. The Bravdo et al. papers are the most cited in reference to the 5–10 Y:PW (i.e., fifty to seventy-five times, which is high in the world of viticulture).

74. Kliewer and Dokoozlian 2005 cite Bravdo et al. 1984 and 1985a when stating that 5–10 is considered good. Bravdo et al. had no values above 10 for Cabernet Sauvignon. For Carignane, there were only two data with yield:pruning weight lower than 10, occurring in 1977 and 1981. In 1981, the wine quality was highest for the low yield:pruning weight; but, in 1977, the wine quality was the lowest for the low yield:pruning weight. Thus, the conclusion that values above 10 are to be avoided is almost baseless from those data. Bravdo et al. 1985a cited a few earlier papers in referring to a Y:PW range for good wine quality, but those included Kliewer et al. 1983, in which 5 of 6 Y:PW values were below 5, and lower quality was not attributable to them. Keller et al. 2005, 2008 cited Smart et al. 1990 to say that 5–10 is good; Smart et al. 1990 (the printed version of a grapevine canopy management talk that appeared in the *South African Journal of Viticulture and Enology*) said 6–10 was the objective, citing the Bravdo et al. 1985a Cabernet Sauvignon paper and a Reynolds 1989 Riesling paper for the 10, and no reference for the 6. What Reynolds says is this: "Crop loads below 10 kg of fruit per kilogram of cane prunings are necessary to achieve adequate fruit maturity under Okanagan conditions." Once again, sugar = quality, and that was the basis for quality assessment. Vasconcelos and Castagnoli 2000 state, "The Ravaz index . . . should remain between 5 and 7" (citing Ravaz 1903, but there were no F/V data or discussion in that paper).

75. Chapman 2004.

76. See example in Lider et al. 1973; similar results are reported in Lider et al. publications in 1967, 1968, and 1975, and in other studies.

77. Kliewer et al. 1983, citing Freeman et al. 1979.

78. As in Freeman et al. 1979 (from 80 to 160 buds/vine) and May et al. 1973 (from 190 to 250 buds/vine).

79. Ravaz 1909, 288.

80. Peterlunger et al. 2002.

81. Nuzzo and Matthews 2006.

82. Freeman et al. 1979: five-year means.

83. Most studies don't present the two parameters from which a correlation could be calculated.

84. Kliewer and Fuller 1973. More recently, Main and Morris 2004: "Leaf-removal treatments had no effect on cluster number, cluster weight, berry weight, yield, pruning weight, or Ravaz index in any year. This is consistent with reports on other grape species and cultivars (several studies cited)."

85. Pruning weight was poorly or negatively related to leaf area in Keller et al. 2005, 2008; Nuzzo and Matthews 2006; Bell and Robson 1999; Palliotti et al. 2011; O'Daniel et al. 2012. Table 1 in Kliewer and Dokoozlian 2005 includes points with a good relation between leaf area and pruning weight, but Table 2 has 4 points with a poor relation; Table 3 has a fair relation, and Tables 4 and 5 have poor relations.

86. Dokoozlian and Kliewer 1995; Etchebarne et al. 2010.

87. Cavallo et al. 2001; see also Poni et al. 2007.

88. Kliewer 1970; Kliewer and Antcliff 1970; Kliewer and Weaver 1971; Kliewer and Ough 1970.

89. Etchebarne et al. 2010.

90. O'Daniel et al. 2012.

91. Reynolds et al. 1994b.

92. Dry et al. 2004.

93. A few examples: Creasy and Creasy 2009; White 2003; Dougherty 2012; Swinchatt and Howell 2004; and Grainger 2009.

94. Zhuang et al. 2014.

95. Tomasi et al. 2014, 13.

96. Santesteban et al. 2010, 51. Echoing the sentiment, Peter Dry and coauthors noted in their essay on vine balance: "One cannot find a clear and accepted definition of the term in the literature"; Dry et al. 2004.

97. Reynolds et al. 1994b.

98. Palliotti 2012.

99. Quote from Amerine and Roessler 1976, 2; viticulture consultant and author Mark Greenspan (2011, 68) says of vine balance, "I know it when I see it."

CHAPTER 3. CRITICAL RIPENING PERIOD AND THE STRESSED VINE

1. For cotton, e.g., see Mackie 1912.

2. Jackson 2008, 212.

3. Birnbaum and Cato 2000, 48; Bastianich and Lynch 2005, 99, 142, 216, 354; Clarke 1988, 15; 2009, 206.

4. Jones and Davis 2000.

5. Winkler 1962, 119; Winkler et al. 1974, 143. Winkler's first comments on heat summation came in articles in the *Wine Review* 1937 and 1938.

6. It takes nine Fahrenheit degree days to make five Celsius degree days.

7. Amerine and Winkler 1944.

8. Gately 2008, 440.

9. Réaumur 1735. See the review by Wang 1960 for a description of Réaumur's work.

10. Wang 1960.

11. Bioletti 1915, 81–88; see also Bioletti 1929, 7–12.

12. Amerine and Winkler 1944.

13. Apparently Huglin, who worked at a station in Colmar, France, developed a comprehensive data set from field trials similar to that published by Amerine and Winkler 1944 and 1963; Hans Schultz (University of Geisenheim): "There is a very good, classical trial, which I always wanted to analyze myself, but never was able to receive the data. Pierre Huglin himself planted grapevine trials in different European countries 40 years ago, from which he also derived his planting recommendations. This trial has been analyzed by Dr. Norbert Becker at the Viticultural Institute in Freiburg" (personal communication). For more on the Huglin Index, see the work of Eric Duchêne (Duchêne and Schneider 2005) and of Greg Jones (Jones and Davis 2000).

14. Tonietto and Carbonneau 2004.

15. McIntyre et al. 1987, 1982.

16. From as early as Gasparin (1844) and Coutagne (1882), both cited in Abbe 1905; and with respect to plant metabolism as described by Livingston and Livingston 1913.

17. Wang 1960.

18. The same goes for other applications of degree days, e.g., to insect development, where the use has exceeded its verification as well; see, e.g., Pruess 1983.

19. McIntyre et al. 1987; this important study has been virtually ignored by both the wine press and academia, having been cited only thirteen times according to the ISI Web of Science database.

20. Coombe 1973, 1980; Due et al. 1993.

21. Schultz 1993. This is based on an average daily temperature of 18°C.

22. Hale and Buttrose 1974.

23. Sadras and Petrie 2011.

24. Hale and Buttrose 1974; Poudel et al. 2009.

25. Duchêne and Schneider 2005.

26. Interestingly, a recent analysis of published studies found that regardless of the seasonal weather, veraison in Cabernet Sauvignon occurs at approximately 65 DAA, and sugar accumulation in the berry ceases at approximately 110 DAA, nicely coincident with the Bordeaux rule of thumb for harvest. See M. A. Matthews, "The End of Ripening" (paper presented at the 2014 National Conference of the American Society for Enology and Viticulture, Austin, TX).

27. Some hold to the idea that faster ripening is better, because the fruit will ripen more before weather brings a close to the season.

28. See, e.g., Robinson 1999, 580; Clarke 2009, 206; and http://www.jancis-robinson.com/learn/grape-varieties/white/chardonnay; http://www.centralotag-opinot.co.nz/news/13_New%20Zealand%E2%80%99s%20pinot%20 noirs%20gaining%20a%20solid%20reputation.html; http://www.reuters.com /article/2014/08/22/usa-california-wine-idUSL2N0QR2JX20140822.

29. Covert 1902; Wetmore 1884.

30. Chassy et al . 2012.

31. Gambetta et al. 2010.

32. Unpublished results of experiments by Mark Krasnow, Mark Matthews, and Ken Shackel, presented by Mark Krasnow at the 2010 National Conference of the American Society for Enology and Viticulture, Seattle, WA.

33. Buttrose et al. 1971; Mori et al. 2007.

34. Girard et al. 2002.

35. Marais 1983; Terrier et al. 1972; Du Plessis 1984.

36. Keller et al. 1998.

37. Downey et al. 2004.

38. Butler 1910, 117: "The California vine disease [aka Pierce's disease] usually develops only when the vines are ripening their fruit, that is, at the most critical period of their yearly development." Weigend 1954 also refers to the "two critical months" in discussing quality of vintages in the first half of the twentieth century.

39. Dering 2007: "during the critical ripening period of early August through early November for winegrapes in Washington State."

40. Clarke and Rand 2007, 15.

41. Wilson 1998, 38.

42. Jones 1988, 24–25; Berlow 1982.

43. The practice may have also started independently in the Tokaj region of Hungary 150 years earlier.

44. Hancock 2009.

45. Hussman 1888, 15–24. Hence, the California wine laws allowing the addition of water.

46. Lance Cutler, "Industry Roundtable: Red Wine Fermentation," *Wine Business Monthly,* April 15, 2007, www.winebusiness.com/wbm/; http://www .winepros.org/wine101/viniculture.htm; http://chateauhetsakais.com/assessinggra pematurity/.

47. Goode 2005 equates physiological maturity with "phenolic maturity."

48. Christian Moueix (founder of Dominus), as quoted in Swinchatt and Howell 2004, 150 (emphasis mine).

49. Hall and Morison 1906, 474. Soon afterward (1918), physiological maturity was introduced in human development to distinguish mental vs. physical (physiological) maturity, the latter occurring earlier than the former.

50. Miller 1946.

51. Watada et al. 1984 (emphasis mine). The "horticultural maturity" definition from the ASHS is very similar to the term "legal maturity," which was distinguished from physiological maturity by Miller in 1946.

52. Although in some cases the seed can be "rescued" and germinated earlier.

53. Courchet 1897, 1:142.

54. Pacottet 1905, 37.

55. Robinet 1877, 3.

56. Franz 1994; additional examples in MacNeil 1998.

57. Coates 1997; and Parker 1990, 2003.

58. California Grape Crush Reports, available from USDA National Agricultural Statistics Service.

59. See Cadot et al. 2012: "Astringency, bitterness, color intensity and alcohol significantly increased with ripening and astringency."

60. MacNeil 2001, 11.

61. No meaning can be gleaned from the viticulture literature, where references to physiological maturity began to appear at the same time as in the popular press. In the earliest use in the viticulture literature, Amerine and Roessler 1958 state that physiological maturity is not sugar, pH, or percent acidity, rather than what it is. I located eight instances where "physiological maturity" appears in peer-reviewed viticulture literature, five of which use the term with no explicit or contextual meaning. Curiously, six of those papers are Spanish groups, but the meaning is variable (Brix/acid ratio, concentration of fatty acids) or completely undefined. In one case, physiological maturity was indicated as coming three weeks after "commercial maturity" of Flame Seedless—the inverse of the rest of horticulture. Olle et al. 2011 and Guyot and Dupraz 2004 apparently use "physiological maturity" synonymously with Brix.

62. Gawel 1998.

63. Sun et al. 1999.

64. Vidal et al. 2004.

65. Fowler 1994, 828.

66. Matthews and Anderson 1988; Castellarin et al. 2007; Bucchetti et al. 2011.

67. Koch et al. 2012; Ryona et al. 2008.

68. E.g., Matthews and Anderson 1988; Castellarin et al. 2007; Dokoozlian and Kliewer 1996.

69. Kalua and Boss 2010.

70. Passmore 2009; Nowak and Wichman 2005, 39.

71. West 2003.

72. Scott-Moncrieff 1868, 30–40; no mention of irrigation for vineyards in Rosenthal 1990; and the only mention of irrigation of vineyards was for phylloxera control in Vincent 1882.

73. Amerine and Joslyn 1970, 72.

74. An irrigation ban remained a component of traditional appellation rules up to 2006; see Jancis Robinson, "Irrigation Now Official in France," http://www.jancisrobinson.com/articles/20070411.html.

75. Johnson 1983, 15.

76. Parker 2008.

77. Hinton and United States Department of Agriculture 1890, 18; Pinney 1989.

78. California Board of State Viticultural Commissioners 1884, 52–70.

79. Henzell 2007, 261.

80. Freeman et al. 1980 in Australia.

81. Compare growth of leaves, internodes, and tendrils at low water status in Schultz and Matthews 1988 with the growth of berries at low water status in Roby and Matthews 2004.

82. MacNeil 2001, 18; see Coombe and Monk 1979 for the late season irrigation study.

83. E.g., "There was a dilution of the intensity of flavor in the grapes as heavy rains drenched the vineyards" (Parker and Rovani 2002, 170).

84. Unpublished data courtesy of Ken Shackel and Brian Bohr Jiun Chen.

85. Matthews et al. 1990.

86. Nitrogen: Ough and Lee 1981; berry color, anthocyanins: Soubeyrand et al. 2014.

87. Keller 2005.

88. Smart et al. 1990.

89. Dokoozlian 1990. This work is also discussed in the context of berry size in chapter 1. Similar results were obtained in field trials with Cabernet Sauvignon by Bergqvist et al. 2001.

90. Koch et al. 2012.

91. Nowak and Wichman 2005, 39.

92. Simon 2003, 85.

93. Matthews and Boyer 1984.

94. E.g., Schultz 2003.

95. Schultz 2000.

96. Allen and Schuster 2004.

97. Parkes and Wollowicz 1870; excellent wine for a study of physiological consequences of wine consumption.

98. Royal Society of Arts 1764, 89–90. Delayed harvest practice is not as new as it may seem, as there are scattered references to waiting for fruit to shrivel dating way back. In one of the earliest testimonies to the benefits of longer hang times, a report claims that Muscats in Orleans, France, were sometimes allowed to hang until November, even allowing for some rot, in order to access the most flavor and highest price. But these fruit were nevertheless unlikely to have been able to reach the high Brix of today.

99. Jim Boswell, http://www.avoidbadwine.com/is-2011-california-wine-really-that-bad/.

100. Brook et al. 2011, 122.

101. Heymann et al. 2013.

CHAPTER 4. THE TERROIR EXPLANATION

1. In French-speaking Canada there is a terroir school of literature, and sundry other applications of terroir—sometimes rather pejorative; challenges in translating these from French to English are reviewed in Tomasik 2001.

2. Pape 2007, x.

3. Bruno Prats, at that time the proprietor of Château Cos d'Estournel, quoted in Halliday 2006, vi. Prats was one side of a public dispute thirty years ago about the role of soil in wine flavor, the other side being Bill Jekel of Jekel Vineyards in Monterey County, California. See, e.g., Jekel and Prats 1983; Chroman 1983.

4. Kramer 1990, 39.

5. Langewiesche 2000, 60.

6. Michel et al. 2011. This is the original work on which Google Ngrams are based.

7. Falcetti's 1994 paper, "Terroir: What Is 'Terroir'? Why Study It? Why Teach It?" (Le terroir: Qu'est-ce qu'un terroir? Pourquoi l'étudier? Pourquoi l'enseigner?) was published in the Bulletin of the International Organization of Vine and Wine (OIV). The OIV is a political organization established in 1927 for the promotion and protection of the wine industry. It is based in Paris, where its offices are afforded embassy status. The OIV has approximately forty member nations; the United States terminated its membership in the OIV as of June 2000. In addition to its member governments, the OIV maintains close contact with such organizations as the Food and Agriculture Organization of the United Nations, the World Health Organization, the World Intellectual Property Organization, the General Agreement on Tariffs and Trade, the European Economic Community, and many other specialized international bodies. The OIV publishes the Bulletin de l'OIV, a quasi-scientific journal that describes itself as an international technical review dealing with viticulture, oenology, economy, and "vitiviniculture" law. Carey et al. 2002 agree with Falcetti, offering an almost identical descriptionof the origins of terroir.

8. Aristotle, *On Youth and Old Age, On Life and Death, On Breathing,* p. 1098 (trans. Ross). Aristotle assumed that all nutrition was acquired by the roots, and that by some process in the soil it was prepared for uptake (as in a stomach). This "food" was complete, and that meant that plants had no excretions, and furthermore, that it contained all things the plant needed right down to the "savours of the fruits." Aristotle, p. 745 (trans. Ross), on plant nutrition: "That part where food enters we call upper, considering it by itself and not relatively to the surrounding universe, while downward is that part by which the primary excrement is discharged. Plants are the reverse of animals in this respect. To man in particular among the animals, on account of his erect stature, belongs the characteristic of having his upper parts pointing upwards in the sense in which that applies to the universe, while in the others these are in an intermediate position. But in plants, owing to their being stationary and drawing their sustenance from the ground, the upper part must always be down; for there is a correspondence between the roots in a plant and what is called the mouth in animals, by means of which they take in their food, whether the source of supply be the earth or each other's bodies."

9. Galen, *On Diseases and Symptoms,* p. 227 (trans. Johnston).

10. Galen, quoted in Prioreschi 1998, 413.

11. Nuland 2001, 55–63.

12. Lee 1887.

13. Ibid.

14. Zirkle 1969.

15. Lachiver 1988, 62.

16. From the website of Stelo Italian wine, http://www.stcloitalianwine .com/.

17. Johnson 1989, 131.

18. Bohmrich 1996.

19. Berman 2000.

20. Berman 1996, 22; Duby 1992, 33.

21. Berman 1986.

22. Rodier 1949, 74. Clos de Vougeot is now 124 acres and has more than eighty owners; Unwin 1996, 178.

23. Roux 2009, 11.

24. Conversi: "Lay brothers in a religious order. The term was originally applied to those who, in adult life, voluntarily renounced the world and entered a religious order to do penance and to lead a life of greater perfection. . . . Among the conversi there were not seldom those who were either entirely illiterate, or who in the world had led a life of public scandal, or had been notorious criminals, and while on the one hand it was unjust that such should be debarred from the means of doing penance in the cloister and from the other benefits of the religious life, they were at the same time hardly to be considered fit subjects for the reception of Sacred orders." (*Catholic Encyclopedia,* s.v.).

25. Unwin 1996, 254–57.

26. Although wine competitions with expert judges date back at least to the world's fairs of the nineteenth century, the next mention of wine scores that I have found appears in research publications by Maynard Amerine at UC Davis

in the 1950s, which would make the monks 500 to 700 years ahead of the curve on that account.

27. Pitte and DeBevoise 2008; Duby 1998; Dion 1959. I also found no description of terroir, clos, or Pinot noir in the context of the rise of the Cistercians in Braudel 1992; Bouchard 1991; Woods 2005; Anderson 2006; or Tobin 1996. Tim Unwin mentions the role of the medieval church in European viticulture, but does not refer to development of the "terroir" idea nor a Cistercian role in it. Alfonso 1991 argued that the Cistercians were very much like other feudal landlords—using forced labor and indentured servants to farm the land. Duby 1992, 239, said that the Cistercians were hard bargainers and perhaps brutes.

28. Coulton 1928, 2.

29. Darcy-Bertuletti 2008 (Yvette Darcy-Bertuletti lays out the complexity).

30. Skinner 1826.

31. Weigend 1954.

32. Coates 1995.

33. Coates 2004, 42–45 (citing research by Pijassou in Médoc).

34. Hailman 2006, 153.

35. Paguierre 1828.

36. Unwin 2000.

37. Busby 1838, 133.

38. Vaudour 2001.

39. Furetière 1690.

40. Paguierre 1828; Allen 1853.

41. Jullien 1866, 232.

42. Ibid., 502. "Les plants tires du Medoc fournissent une faible quantité de vin assez bon, qui, malgré un gout particulier de terroir, ressemble plus au vin de Bordeaux qu'a tout autre." (502; Plants harvested from Medoc provide a small amount of wine of quality, which, despite a particular local flavor, is more like Bordeaux wine than any other.)

43. Chaptal 1801, 343.

44. "The Manufacture of Brandy at Cognac," *The Times,* October 21, 1889.

45. Grazzi-Soncini and Bioletti 1892, 46.

46. A. Elne, "A Forgotten City" (From a correspondent), *The Times,* July 9, 1928.

47. Knight 1837; Sachs et al. 1890, 517. Sachs et al. say the spongiole came from the eighteenth-century French botanist De Candolle as "an unfortunate invention of his own."

48. Garrett 2003, 63.

49. Mayr 2004, 21–38.

50. Hales 1727, 376.

51. Morley 1852, 311.

52. Van Helmont was an interesting character. Nuland 2001 seems to think him more of a crackpot: "Jean Baptiste van Helmont (Belgian) was a perennially troubled and confused man. . . . His formulations, and in fact his life, are somewhat like a pendulum of uncertainty, swinging back and forth between science and mysticism." In addition to publishing "The Magnetic Cure of

Wounds," van Helmont believed that he had converted mercury to gold, and that the stomach was the key organ, containing the soul, a motivating spiritual quality, and the mind. Nevertheless, van Helmont was a leading force in the development of chemistry.

53. Van Helmont's work on plant growth was published after his death. His experiment is held up as an important example of good early science in *Elegant Solutions: Ten Beautiful Experiments in Chemistry,* by Philip Ball (Ball 2005).

54. Leicester and Klickstein 1963.

55. Hershey 2003.

56. Mariotte 1740, 121–47.

57. Stroup 1990, 148.

58. Sachs et al. 1890, 463–64.

59. Tull 1762, 20.

60. Albrecht Thaer (1752–1828) is credited with the Humus Theory of plant nutrition.

61. In the mid-1600s, John Glauber developed the first chemical fertilizer, a combination of saltpeter, lime, phosphoric acid, and potash.

62. Liebig 1847. Carl Sprengel was one of Thaer's former collaborators. In an effort to acknowledge Sprengel's contribution, some now refer to the Sprengel-Liebig Law of the Minimum.

63. Priestley 1775, 182.

64. Priestley 1777, 250.

65. Franklin, as quoted in Sparks 1840, 414.

66. Priestley also discovered how to make soda water, and this discovery led to the soda industry. Natural soda water was thought to have many medicinal purposes.

67. As quoted in Magiels 2010, 64.

68. Haeger 2004. John Haeger does a nice job discussing the fuzzy origins of Pinot noir in Burgundy.

69. Gholami et al. 1995; Koch et al. 2010. In both studies, when clusters of a flavor-producing variety were grafted onto the shoot of a non flavor-producing variety, the ripened grapes contained the flavor (monoterpenes or MIBP), and when clusters of a non flavor-producing variety were grafted onto shoots of a flavor-producing variety, the ripened grapes did not develop the flavor. These results show that it is the genotype of the cluster that determines its characteristic flavor traits, and not the shoot on which it grows.

70. Redding 1851, 218.

71. Seguin 1975; Duteau et al. 1981; Seguin 1986.

72. Duteau and Seguin 1973.

73. Debbie Elliot-Fisk, physical geographer at UC Davis, surveyed soils in Napa and Bordeaux in the 1990s and found that soils are more varied in Napa Valley than Bordeaux, because of the complex geological history of the Napa region; however, the wines produced in Napa are not recognized as more diverse.

74. Wilson 1998; Pomerol 1989, 3–15, 91, 247.

75. Lafforgue and Chappaz 1947, 52, as cited in Weigend 1954; also Seguin 1986 noted the failure of iron or other mineral theories as a basis for terroir to hold up to further study.

76. Guyot 2010, 48, originally published in 1861. Guyot introduced system of long or "cane-pruning" of vines for trellises. The Guyot training system is extensively used throughout vineyards in Europe.

77. Amerine and Joslyn 1970, 33.

78. Olken 2012; Wilson 2011; Monroe 2007.

79. Redding 1851, 218.

80. Adapted from the Soil Science Society of America definition of a soil at https://www.soils.org/publications/soils-glossary#.

81. Hobart M. King's geology.com website.

82. Pomerol and France Bureau de recherches géologiques et minières 1989; Haynes 1999.

83. I refer especially to the archaic soil classification system used today.

84. Gale 2004.

85. Johnson 1999.

86. Logan 2007, 20.

87. Ibid., 64.

88. Zimmerman 2009.

89. Nitrogen, phosphorus, potassium, calcium, magnesium, sulfur, iron, zinc, manganese, copper, boron, molybdenum, chlorine, and nickel.

90. Increased nitrogen supply in the soil usually results in increased nitrogen in the fruit, and many flavor compounds originate in the fruit from nitrogenous molecules. For example, Ough and Lee 1981 show increased fruity esters in wines made from vines with higher rates of N fertilizer applications. Red color synthesis responds to the fruit nitrogen status as well, where high N can reduce color synthesis.

91. Webster et al. 1993; Ough and Lee 1981.

92. Sumby et al. 2010.

93. Jullien 1866, 21.

94. Bormann 1957.

95. Heap and Newman 1980.

96. Farina et al. 2005; Capone et al. 2012.

97. The black walnut tree remains a good example of allelopathy, and the growth inhibitor it produces has been isolated and identified.

98. Jullien 1866, 158 and 232.

99. Gerbaux et al. 2000.

100. Rosenthal 2008. Mr. Rosenthal likes the contribution by Brett. See also a discussion of the positive side of Brett in http://culinarywineandfoodmatching .blogspot.com/2009/10/positive-taste-of-brett-in-wines-and.html.

101. Henick-Kling et al. 2000.

102. It has been speculated that it is the scent that camels follow to find water.

103. Darriet et al. 2000.

104. Wenke et al. 2010.

105. Simpson 2005.

106. Demossier 1997.

107. Ibid.

108. Whalen 2007b.

109. Whalen 2007a.

110. Ibid.; Labbé 1938.

111. Gough 1998.

112. Whalen 2009.

113. Boisard 2003.

114. Ulin 1987.

115. Demossier 1997, 56.

116. Féret et al. 1962.

117. Debuigne 1970.

118. Ramsey 1929, 238, The joke refers to Wittgenstein's expertise at whistling.

119. Leading ultimately to both Australia and the United States overtaking France in wine sales in Britain, along with increased competition from Italy, Chile, and South Africa, among others. "America Wine Overtakes France in British Drinkers' Affection," *The Telegraph*, December 15, 2008.

120. Blind tasting—the judges did not know the identity of the wines they tasted.

121. Taber's book *Judgment of Paris: California vs. France and the Historic 1976 Paris Tasting* is a good read for wine lovers, and led to the movie "Bottle Shock." The original magazine article is still available at http://www.time.com /time/magazine/article/0,9171,947719,00.html.

122. Joanne DePuy, who arranged the tour: http://www.napawinelockers .com/depuy.html.

123. Winiarski 1991.

124. Some authors would apparently prefer to forget about the famous tasting; e.g., Whalen 2010. On the other hand, the Americans continue to celebrate its anniversaries: http://blogs.usda.gov/2011/06/22/california-wine-industry-celebrates -35th-anniversary-of-the-%E2%80%98judgment-in-paris%E2%80%99. In an effort to refute the Judgment results, some point to Lindley 2006, which argues it is difficult to attribute statistically significant differences to the wine scores from the Judgment of Paris. However, the lack of clear differences is probably the main conclusion to be drawn.

125. Ali et al. 2005.

126. Quandt 2007b, 222.

127. R. E. Quandt, http://www.liquidasset.com/report75.html.

128. The relative ease of committing wine fraud by claiming a false source of grapes makes the same argument: e.g., see http://www.nytimes.com/2015/01/25 /business/in-vino-veritas-in-napa-deceit.html?emc = eta1&_r = 2.

129. Tanzer 2006, 16–17.

130. Peynaud 1996, 303.

131. Prial 1989. Prial wrote the *New York Times* wine column for twenty-five years. The late Alexis Lichine (who owned Château Prieuré-Lichine and a share of Château Lascombes in the Médoc) spent much of his professional life arguing for a revision of the 1855 classification of the wines of Bordeaux. He didn't succeed. In fact, there has been but one change recorded.

132. Peynaud 1996, 303–5.

133. Jackson 2000, 484–86.

134. Croidieu 2011, 303.

135. Meloni and Swinnen 2012. The history of European wine laws has been well reviewed also by Dev Gangjee (2012), and to some extent by Warren Moran (1993). They all report that the laws produce high rents for the land.

136. Dion 1959, VIII.

137. Gangjee 2012, 115.

138. Josling 2006.

139. Broude 2005.

140. Moran 1993.

141. Barham 2003; Ulin 1995.

142. Gade 2004.

143. Mueller and Sumner 2006.

144. Paul 2005.

145. Castle 2012. "E.U. rules derail English-Argentine wine project; Importation of grapes renders British bottles unsuitable for labeling." In this article, a UK winemaker learned he could not label his Malbec as Malbec or even as "wine" because it was made from grapes produced outside the EU.

146. Broude 2005.

147. A brief history of *typicité* is provided in the 2013 doctoral dissertation of Armin Schuttler (Schuttler 2013). Whereas in French, *typicité* is used to associate a putative wine attribute with terroir; in German the term has usually been used as a synonym for "varietal character."

148. At the initial meeting of an advanced viticulture class on terroir, a student walked out never to return, because I said I was going to focus on terroir to the exclusion of typicité.

149. Transition to the new positive terroir: "a wine has flavors reflecting a combination of soil, slope, and climate" (Joseph 1996, 69); "the characteristic taste and flavor imparted to a wine by the environment in which it is produced" (*Oxford English Dictionary,* online ed.).

150. Robinson 2002b.

151. Maltman 2008.

152. See note 7; Josling 2006.

153. Quotations: Robinson 1999, 700; Laville 1990; Carey et al. 2002.

154. See, e.g., Nick Bulleid, "Pinot noir: Terroir or Intervention," *Wine Industry Journal* 14 (1999):26–28.

155. Peynaud 1996.

156. Pereira et al. 2005.

157. See *Journal of the Department of Agriculture, Victoria* 27 (1930):654. A more neutral description of *goût de terroir* as an earthy taste appears in Lichine 1967, 257.

158. Prial 1979.

159. *ABC des vins* (Paris: Larousse, 1995), 219: "Terroir : Ensemble des sols, des sous sols, de l'exposition et de l'environnement, qui déterminent le caractère d'un vin."

160. The approaches referred to here are for the development of rules for boundaries or zoning to regulate what goes on a wine label.

161. Ribéreau-Gayon 2000, 270.

162. http://knowledge.wharton.upenn.edu/article/the-future-of-french-wine-overcoming-terroirisme-and-stagnation/.

163. Pereira et al. 2005.

164. Van Leeuwen et al. 2004.

165. Fischer et al. 1999.

166. "Grower and Cellar News," *Wine Business Monthly,* January 20, 2000.

167. Morris 2011.

168. Morlat and Bodin 2006.

169. See Pliny, *Natural History* 4.8.59–62 (trans. Rackham): "Who can doubt, however, that some kinds of wine are more agreeable than others, or who does not know that one of two wines from the same vat can be superior to the other, surpassing its relation either owing to its cask or from some accidental circumstance? And consequently each man will appoint himself judge of the question which wine heads the list."

170. Guy 2003.

171. Durand 1992, 797; cited and translated by Brazeau 2009, 29.

172. Barthes 1970; cited and translated by Brazeau 2009, 29.

173. Paraphrased from Gross and Levitt 1997, 71.

174. Adler and Weingarten 2005.

175. Wilson 1998, 118. Despite her contention that it is all too complicated to ever understand, Bize-Leroy says, "I remain convinced however, that the fundamental character of each wine depends on the nature of the subsoil." Some insight into her predisposition might be gained from this notice on her winery webpage: "Les vignes du Domaine LEROY sont cultivées en Bio-dynamie depuis 1989." (We've been biodynamic since 1989.)

176. Kunzig 1999.

177. Ibid.

178. Kramer 1990, 39–40.

179. Fourcade 2012.

180. Paul Lukacs, http://zesterdaily.com/drinking/dont-get-hung-up-on-terroir-its-a-myth/

EPILOGUE

1. It may also surprise some that both Bioletti in the early 1900s and Nelson Shaulis at Cornell University in the mid-1900s argued that yield could be increased in many vineyards with no loss of quality. In fact, increasing yield while maintaining quality was the stated goal of Shaulis's work on "balanced pruning."

2. Another indication of the isolation of wine producers from the plant world.

3. Krasnow et al. 2008.

4. Mueller and Sumner 2006.

5. Following the aphorism "It is better to be vaguely right than exactly wrong," in *Logic: Deductive and Inductive* (1920) by British philosopher Carveth Read.

6. Sadras and Moran 2012; Sadras et al. 2013.

7. E.g., Hughson and Boakes, 2002.

8. Frankfurt 2009, 63.

9. Quandt 2007a; he argues that wine professionals neither agree on wine quality nor actually convey information in their wine descriptions.

10. Matthews and Anderson 1988.

11. Moran 2001.

12. Extrinsic factors are heavily influential; and wine tasting professionals tend to apply extrinsic explanations to sense experience—e.g., complexity; see Parr et al. 2011.

13. "Fragmentation and Over-Regulation Are Undermining the Wine Producers of the Old World," *The Economist*, December 18, 1999.

14. Ashenfelter et al. 2013.

15. Pinker 1997, 555.

16. According to Wikipedia, anthroposophy is "a philosophy that postulates the existence of an objective, intellectually comprehensible spiritual world accessible to direct experience through inner development. More specifically, it aims to develop faculties of perceptive imagination, inspiration, and intuition through the cultivation of a form of thinking independent of sensory experience."

17. Steiner 1958, 5.

18. Biodynamics is more about a human spiritual connection to farming than a predictor of wine quality. In this sense, it fits well into today's new morality-based eating. See, for example, Stein and Nemeroff 1995.

19. Pinker 1997, ix.

20. Adapted from Sir Peter Medawar, winner of the 1960 Nobel Prize in Medicine; Medawar 1984, 15–17.

21. Boulton 2012.

22. See work by Julian Alston on return on investment in agricultural research; e.g., Alston et al. 2011.

23. Black 2013.

24. Viticulture researcher Mike Anderson, personal communication.

25. Myles et al. 2011.

26. This is a generalization; there are some successful winegrape breeding programs, primarily in the most challenging (cool/cold) environments.

27. Gergaud and Ginsburgh 2008.

28. Plassmann et al. 2008.

29. McCoy 2005.

30. See, e.g., the social media site winesearcher.com and mobile apps "delectable" and "vivino"; Hanni 2013.

References

Abbe, C. 1905. *A First Report on the Relations between Climates and Crops.* U.S. Weather Bureau Bulletin. Washington, DC: Government Printing Office.

Adler, J., and T. Weingarten. 2005. "The Taste of the Earth." *Newsweek,* February 27, 2005.

Alfonso, I. 1991. "Cistercians and Feudalism." *Past & Present* (133):3–30.

Ali, H.H., et al. 2005. "The Impact of Gurus : Parker Grades and en primeur Wine Prices." Document de travail de l'INRA, Laboratoire d'Économie Appliquée, INRA, Paris.

Allen, J. F. 1853. *A Practical Treatise on the Culture and Treatment of the Grape Vine: Embracing Its History, with Directions for Its Treatment, in the United States of America, in the Open Air, and under Glass Structures, with and without Artificial Heat.* 3rd ed. New York: C.M. Saxton.

Allen, S. J., and E. W. Schuster. 2004. "Controlling the Risk for an Agricultural Harvest." *Manufacturing & Service Operations Management* 6 (3):225–36.

Alonso-Cantabrana, H., et al. 2007. "Common Regulatory Networks in Leaf and Fruit Patterning Revealed by Mutations in the Arabidopsis ASYMMETRIC LEAVES1 Gene." *Development* 134 (14):2663–71. doi: 10.1242/Dev.02864.

Alston, J.M., et al. 2011. "The Economic Returns to U.S. Public Agricultural Research." *American Journal of Agricultural Economics* 93 (5):1257–77.

Amerine, M.A., et al. 1972. *The Technology of Wine Making.* 3rd ed. Westport, CT: AVI Publishing.

Amerine, M.A., and M.A. Joslyn. 1970. *Table Wines: The Technology of Their Production.* Berkeley: University of California Press.

Amerine, M.A., and E.B. Roessler. 1952. "Techniques and Problems in the Organoleptic Examination of Wines." *Proc. Am. Soc. Enol.* 9:97–115.

———. 1958. "Methods of Determining Field Maturity of Grapes." *American Journal of Enology and Viticulture* 9:37–40.

———. 1976. *Wines: Their Sensory Evaluation.* San Francisco: W.H. Freeman.

Amerine, M.A., and A.J. Winkler. 1944. "Composition and Quality of Musts and Wines of California Grapes." *Hilgardia* 15:493–673.

———. 1963. "California Wine Grapes." *California Agricultural Experiment Station Bulletin* 794.

Anderson, D. 2006. *Monks and Monasteries in the Middle Ages.* World Almanac Library of the Middle Ages. Milwaukee: World Almanac Library.

Anderson, M.M., et al. 2008. "Viticultural Evaluation of French and California Chardonnay Clones Grown for Production of Sparkling Wine." *American Journal of Enology and Viticulture* 59 (1):73–77.

Andreski, S. 1972. *Social Sciences as Sorcery.* London: Deutsch.

Aradhya, M.K., et al. 2003. "Genetic Structure and Differentiation in Cultivated Grape, Vitis vinifera L." *Genetical Research* 81 (3):179–92. doi: 10.1017/S0016672303006177.

Archer, E., and H.C. Strauss. 1990. "The Effect of Vine Spacing on Some Physiological Aspects of Vitis vinifera L. (cv. Pinot noir)." *South African Journal of Enology and Viticulture* 11 (2):76–87.

———. 1991. "The Effect of Vine Spacing on the Vegetative and Reproductive Performance of Vitis vinifera L. (cv. Pinot noir)." *South African Journal of Enology and Viticulture* 12 (2):70–77.

Aristotle. *On Youth and Old Age, On Life and Death, On Breathing.* Translated by G.R.T. Ross. http://classics.mit.edu/Aristotle/youth_old.html.

Ashenfelter, O., et al. 2013. "Wine Tasting: Is "Terroir" a Joke and/or Are Wine Experts Incompetent?" *VOX CEPR's Policy Portal,* http://www.voxeu.org/article/wine-tasting-terroir-joke-andor-are-wine-experts-incompetent?quicktabs_tabbed_recent_articles_block = 1.

Astill, G.G., and J. Langdon. 1997. *Medieval Farming and Technology: The Impact of Agricultural Change in Northwest Europe.* Technology and Change in History. Leiden and New York: Brill.

Bailey, L.H. 1902. *Cyclopedia of American Horticulture.* Edited by Wilhelm Miller. London: Macmillan.

———. 1916. *The Pruning-Manual.* 18th ed., rev. and reset. New York: Macmillan.

Baillie, J.B. 1918. "Anthropomorphism and Truth." *Proceedings of the Aristotelian Society,* n.s., 18:185–223.

Ball, P. 2005. *Elegant Solutions: Ten Beautiful Experiments in Chemistry.* Cambridge: Royal Society of Chemistry.

Barbagallo, M.G., et al. 2011. "Berry Size and Qualitative Characteristics of Vitis vinifera L. cv. Syrah." *South African Journal of Enology and Viticulture* 32 (1):129–36.

Barham, E. 2003. "Translating Terroir: The Global Challenge of French AOC Labeling." *Journal of Rural Studies* 19 (1):127–38. doi: 10.1016/S0743-0167(02)00052-9.

Barnes, J. 2014. *Complete Works of Aristotle.* Vol. 1. Princeton, NJ: Princeton University Press.

Bastianich, J., and D. Lynch. 2005. *Vino italiano: The Regional Wines of Italy.* Rev. and updated ed. New York: Clarkson Potter.

Battistutta, F., et al. 1997. "Influence of Genotype on the Methoxypyrazines Content of Cabernet Sauvignon in Friuli." *Acta Hort. (ISHS)* 526:407–13.

———. 2000. "Influence of Genotype on the Methoxypyrazines Content of Cabernet Sauvignon Cultivated in Friuli." *Acta Hort. (ISHS)* 526:407–14.

Bavaresco, L., et al. 2010. "Nutritional Deficiencies." In *Methodologies and Results in Grapevine Research,* edited by S. Delrot et al. Dordrecht: Springer.

Bell, B. 1980. "Analysis of Viticultural Data by Cumulative Deviations." *Journal of Interdisciplinary History* 10 (4):851–58. doi: 10.2307/203076.

Bell, S.J., and A. Robson. 1999. "Effect of Nitrogen Fertilization on Growth, Canopy Density, and Yield of Vitis vinifera L. cv. Cabernet Sauvignon." *American Journal of Enology and Viticulture* 50 (3):351–58.

Bergqvist, J., et al. 2001. "Sunlight Exposure and Temperature Effects on Berry Growth and Composition of Cabernet Sauvignon and Grenache in the Central San Joaquin Valley of California." *American Journal of Enology and Viticulture* 52 (1):1–7.

Berlow, R.K. 1982. "The 'Disloyal Grape': The Agrarian Crisis of Late Fourteenth-Century Burgundy." *Agricultural History* 56 (2):426–38.

Berman, C.H. 1986. *Medieval Agriculture, the Southern French Countryside, and the Early Cistercians: A Study of Forty-Three Monasteries.* Philadelphia: American Philosophical Society.

———. 1996. "Agriculture." In *Medieval France: An Encyclopedia*, edited by W.W. Libler et al. New York: Garland Publishing.

———. 2000. *The Cistercian Evolution: The Invention of a Religious Order in Twelfth-Century Europe.* The Middle Ages. Philadelphia: University of Pennsylvania Press.

Bernizzoni, F., et al. 2009. "Long-Term Performance of Barbera Grown under Different Training Systems and Within-Row Vine Spacings." *American Journal of Enology and Viticulture* 60 (3):339–48.

Beuerlein, J. 2001. *Soybean Plant Spacing: The Last Frontier.* http://ohioline.osu.edu/agf-fact/0140.html.

Bindon, K.A., et al. 2008. "The Interactive Effect of Pruning Level and Irrigation Strategy on Grape Berry Ripening and Composition in Vitis vinifera L. cv. Shiraz." *South African Journal of Enology and Viticulture* 29 (2):71–78.

Bindon, K.A., and J.A. Kennedy. 2011. "Ripening-Induced Changes in Grape Cell Walls Modify Their Interaction with Tannins." *American Journal of Enology and Viticulture* 62 (3):392a.

Bioletti, F.T. 1897. "Vine Pruning." *California Agricultural Experiment Station Bulletin* 119. http://archive.org/details/vinepruning119biol.

———. 1913. "Vine Pruning in California, Part 1." *California Agricultural Experiment Station Bulletin* 241. http://archive.org/stream/vinepruningincal241biol#page/n21/mode/2up.

———. 1915. "Viticulture on the Pacific Coast." In *Official Report of the Session of the International Congress of Viticulture, San Francisco, CA, July 12–13,* 81–88. San Francisco.

————. 1929. *Elements of Grape Growing in California.* Berkeley: University of California Print Office.

Bioletti, F.T., et al. 1886. *Report of the Viticultural Work during the Seasons 1883–4 and 1884–5 [1885 and 1886, 1887–89, 1887–93].* Sacramento: J.J. Ayres, Superintendent State Printing.

Birnbaum, B.B., and K.D. Cato. 2000. *Geology and Enology of the Temecula Valley, Riverside County, California.* San Diego: San Diego Association of Geologists.

Black, R.E. 2013. "Vino Naturale: Tensions between Nature and Technology in the Glass." In *Wine and Culture: Vineyard to Glass,* edited by R.E. Black and R.C. Ulin. New York: Bloomsbury Publishing.

Bohmrich, R. 1996. "Terroir: Competing Perspectives on the Roles of Soil, Climate, and People." *Journal of Wine Research* 7:33–35.

Boisard, P. 2003. *Camembert: A National Myth.* Berkeley: University of California Press.

Bond, B.J. 2000. "Age-Related Changes in Photosynthesis of Woody Plants." *Trends in Plant Science* 5 (8):349–53. doi: 10.1016/S1360–1385(00)01691–5.

Bonnet, A. 1937. "The Blight" [in French]. *Progrès Agricole et Viticole* 108:325–27.

Bopp, M. 1996. "The Origin of Developmental Physiology of Plants in Germany." *International Journal of Developmental Biology* 40 (1):89–92.

Bormann, F.H. 1957. "Moisture Transfer between Plants through Intertwined Root Systems." *Plant Physiology* 32 (1):48–55. doi: 10.1104/Pp.32.1.48.

Boss, P.K., et al. 2003. "New Insights into Grapevine Flowering." *Functional Plant Biology* 30 (6):593–606.

Boss, P.K., and M.R. Thomas. 2000. "Tendrils, Inflorescences, and Fruitfulness: A Molecular Perspective." *Australian Journal of Grape and Wine Research* 6 (2):168–74. doi: 10.1111/j.1755-0238.2000.tb00176.x.

Boswell, J. "Is 2011 California Wine Really That Bad?" http://www.avoidbadwine.com/is-2011-california-wine-really-that-bad/.

Bouchard, C.B. 1991. *Holy Entrepreneurs : Cistercians, Knights, and Economic Exchange in Twelfth-Century Burgundy.* Ithaca, NY: Cornell University Press.

Boulton, R. 2012. "Discovery, Innovation, and the Role of Research." Presentation for the ASEV Merit Award, at the American Society for Enology and Viticulture's 63rd Annual Meeting, 2012, Portland, Oregon.

Bovey, A. 2005. *Tacuinum sanitatis: An Early Renaissance Guide to Health.* London: Sam Fogg.

Bowen, P., et al. 2011. "Effects of Irrigation and Crop Load on Leaf Gas Exchange and Fruit Composition in Red Winegrapes Grown on a Loamy Sand." *American Journal of Enology and Viticulture* 62 (1):9–22. doi: 10.5344/ajev.2010.10046.

Bramley, R.G.V., and R.P. Hamilton. 2004. "Understanding Variability in Winegrape Production Systems." *Australian Journal of Grape and Wine Research* 10 (1):32–45. doi: 10.1111/j.1755-0238.2004.tb00006.x.

Branas, J., et al. 1946. *Éléments de viticulture générale.* Montpellier: École National d'Agriculture de Montpellier.

Braudel, F. 1992. *Civilization and Capitalism, 15th-18th Century*. 3 vols. Berkeley: University of California Press.

Bravdo, B., et al. 1984. "Effect of Crop Level on Growth, Yield, and Wine Quality of a High-Yielding Carignane Vineyard." *American Journal of Enology and Viticulture* 35 (4):247–52.

———. 1985a. "Effect of Crop Level and Crop Load on Growth, Yield, Must, and Wine Composition, and Quality of Cabernet Sauvignon." *American Journal of Enology and Viticulture* 36 (2):125–31.

———. 1985b. "Effect of Irrigation and Crop Level on Growth, Yield, and Wine Quality of Cabernet Sauvignon." *American Journal of Enology and Viticulture* 36 (2):132–39.

Bravdo, B., and A. Naor. 1996. "Effect of Water Regime on Productivity and Quality of Fruit and Wine." *First ISHS Workshop on Strategies to Optimize Wine Grape Quality* 427:15–26.

Brazeau, B. 2009. *Writing a New France, 1604–1632: Empire and Early Modern French Identity*. Farnham, UK: Ashgate.

Brook, S., et al. 2011. *The Finest Wines of California: A Regional Guide to the Best Producers and Their Wines*. Berkeley: University of California Press.

Broude, T. 2005. "Culture, Trade, and Additional Protection for Geographical Indications." *International Centre for Trade and Sustainable Development, Bridges Monthly* 9:20–22. http://www.iprsonline.org/ictsd/docs/Bridges Monthly9–9GIs.pdf.

Brown, P.H., et al. 1987. "Nickel—a Micronutrient Essential for Higher-Plants." *Plant Physiology* 85 (3):801–3. doi:: 10.1104/Pp.85.3.801.

Bucchetti, B., et al. 2011. "Effect of Water Deficit on Merlot Grape Tannins and Anthocyanins across Four Seasons." *Scientia Horticulturae* 128 (3):297–305. doi: 10.1016/j.scienta.2011.02.003.

Buchanan, R., and N. Longworth. 1852. *The Culture of the Grape, and Wine-Making*. Cincinnati: Moore & Anderson.

Busby, J. 1838. *Journal of a Recent Visit to the Principal Vineyards of Spain and France*. Philadelphia: Sherman and Co. Printers.

Butler, O. 1910. "Observations on the California Vine Disease." *Memoirs of the Torrey Botanical Club* 14:111–53.

Buttrose, M.S., et al. 1971. "Effect of Temperature on Composition of Cabernet Sauvignon Berries." *American Journal of Enology and Viticulture* 22 (2):71–75.

Buttrose, M.S., and C.R. Hale. 1973. "Effect of Temperature on Development of Grapevine Inflorescence after Bud Burst." *American Journal of Enology and Viticulture* 24 (1):14 16.

Cadot, Y., et al. 2012. "Sensory Representation of Typicality of Cabernet franc Wines Related to Phenolic Composition: Impact of Ripening Stage and Maceration Time." *Analytica Chimica Acta* 732:91–99.

Cai, X., and J. Noel. 2013. "California Farmland Valuation: A Hedonic Approach." Paper presented at the Agricultural & Applied Economics Association's 2013 Annual Meeting, August 4–6, 2013, in Washington, DC.

California Agricultural Experiment Station. 1886. *Report of Viticultural Work during the Seasons of 1883–84 and 1884–85.* Sacramento: J.J. Ayres, Superintendent State Printing.

California Board of State Viticultural Commissioners. 1884. *Second Annual Report of the Chief Executive Viticultural Officer . . . 1882–3.* Sacramento: Sacramento State Printer.

Campbell, C. 2006. *The Botanist and the Vintner: How Wine Was Saved for the World.* Chapel Hill, NC: Algonquin Books of Chapel Hill.

Capone, D.L., et al. 2012. "Vineyard and Fermentation Studies to Elucidate the Origin of 1,8-Cineole in Australian Red Wine." *Journal of Agricultural and Food Chemistry* 60 (9):2281–87.

Carey, V.A., et al. 2002. "Natural Terroir Units: What Are They? How Can They Help the Wine Farmer?" *Wineland,* February 2002, 86–88.

Carroll, J. 2004. *Literary Darwinism: Evolution, Human Nature, and Literature.* New York: Routledge.

Castellarin, S.D., et al. 2007. "Water Deficits Accelerate Ripening and Induce Changes in Gene Expression Regulating Flavonoid Biosynthesis in Grape Berries." *Planta* 227 (1):101–12.

———. 2011. "Fruit Ripening in Vitis vinifera: Spatiotemporal Relationships among Turgor, Sugar Accumulation, and Anthocyanin Biosynthesis." *Journal of Experimental Botany* 62 (12):4345–54. doi: 10.1093/Jxb/Err150.

Castle, S. 2012. "When Is a Wine Not a Wine? When European Regulations Say It's Not." *New York Times,* May 29, 2012. http://www.nytimes.com/2012/05/30/world/europe/when-is-a-wine-not-a-wine-when-european-regulations-say-its-not.html?_r = 0.

Cato the Censor. *On Farming = De Agricultura.* Translated and commentary by Andrew Dalby. Totnes, UK: Prospect Books, 1998.

Cavallo, P., et al. 2001. "Ecophysiology and Vine Performance of cv. "Aglianico" under Various Training Systems." *Scientia Horticulturae* 87: 21–32.

Cawthon, D.L., and J.R. Morris. 1982. "Relationship of Seed Number and Maturity to Berry Development, Fruit Maturation, Hormonal Changes, and Uneven Ripening of Concord (Vitis-Labrusca L) Grapes." *Journal of the American Society for Horticultural Science* 107 (6):1097–1104.

Champagnol, F. 1984. *Éléments de physiologie de la vigne et de viticulture générale.* Prades le Lez: L'auteur.

Chapman, D.M. 2004. "Crop Yield and Vine Irrigation Effects on the Sensory Quality of Vitis Vinifera L., Cv. Cabernet Sauvignon and Sangiovese Wine." PhD diss., Food Sciences, University of California, Davis.

Chapman, D.M., et al. 2004a. "Sensory Attributes of Cabernet Sauvignon Wines Made from Vines with Different Crop Yields." *American Journal of Enology and Viticulture* 55 (4):325–34.

———. 2004b. "Yield Effects on 2-Methoxy-3-Isobutylpyrazine Concentration in Cabernet Sauvignon Using a Solid Phase Microextraction Gas Chromatography/Mass Spectrometry Method." *Journal of Agricultural and Food Chemistry* 52 (17):5431–35. doi: 10.1021/Jf0400617.

————. 2005. "Sensory Attributes of Cabernet Sauvignon Wines Made from Vines with Different Water Status." *Australian Journal of Grape and Wine Research* 11 (3):339–47. doi: 10.1111/j.1755-0238.2005.tb00033.x.

Chaptal, J. A. C., et al. 1801. *Traité théorique et pratique sur la culture de la vigne avec l'art de faire le vin, les eaux-de-vie, esprit-de-vin, vinaigres simples et composés.* Paris: Delalain.

Chassy, A. W., et al. 2012. "Tracing Phenolic Biosynthesis in Vitis vinifera via in Situ C-13 Labeling and Liquid Chromatography-Diode-Array Detector-Mass Spectrometer/Mass Spectrometer Detection." *Analytica Chimica Acta* 747:51–57.

Chaussee, J. 2014. "Early Grape Crush Kicks Off California's Winemaking Season." August 21, 2014. http://www.reuters.com/article/2014/08/22/usa-california-wine-idUSL2N0QR2JX20140822.

Chien, M. 2012. "Lessons in Canopy Management from Virginia." http://www.virginiavineyardsassociation.com/wp-content/uploads/2012/06/Canopy-Management-workshop-notes_Mark-Chien.pdf.

Cho, A. 2014. "Elite Violinists Fail to Distinguish Legendary Violins from Modern Fiddles." *Brain & Behavior,* April 7, 2014. http://news.sciencemag.org/brain-behavior/2014/04/elite-violinists-fail-distinguish-legendary-violins-modern-fiddles.

Christensen, P. 1975. "Response of 'Thompson Seedless' Grapevines to the Timing of Preharvest Irrigation Cut-Off." *American Journal of Enology and Viticulture* 26 (4):188–94.

Chroman, N. 1983. "ABOUT WINE—Uncorking a Controversy: Soil-Taste Battle Heats Up between United States and France." *Los Angeles Times,* February 10, 1983, L42.

Clarke, O. 1988. *The Wine Book.* New York: Random House Value Publishing.

————. 2006. *Oz Clarke's Bordeaux: The Wines, the Vineyards, the Winemakers.* Orlando, FL: Harcourt.

————. 2009. *Oz Clarke's Pocket Wine Guide 2010.* New York: Sterling.

Clarke, O., and M. Rand. 2007. *Oz Clarke's Grapes and Wines: The Definitive Guide to the World's Great Grapes and the Wines They Make.* Boston: Harcourt.

Clingeleffer, P. R. 1984. "Production and Growth of Minimal Pruned Sultana Vines." *Vitis* 23 (1):42–54.

Clingeleffer, P. R., et al. 2001. *Crop Development, Crop Estimation, and Crop Control to Secure Quality and Production of Major Wine Grape Varieties: A National Approach; Final Report to Grape and Wine Research & Development Corporation.* Project CSH 96/1. Victoria/Adelaide: CSIRO.

————. 2002. "Effect of Post-Set, Crop Control on Yield and Wine Quality of Shiraz." In *Proceedings of the Eleventh Australian Wine Industry Technical Conference,* edited by R. J. Blair et al., 84–86. Adelaide, Australia: Australian Wine Technical Conference.

Clingeleffer, P. R., and L. R. Krake. 1992. "Responses of Cabernet franc Grapevines to Minimal Pruning and Virus Infection." *American Journal of Enology and Viticulture* 43 (1):31–37.

Coates, C. 1995. *Grands Vins: The Finest Châteaux of Bordeaux and Their Wines*. Berkeley: University of California Press.

———. 1997. *Côte d'Or: A Celebration of the Great Wines of Burgundy*. Berkeley: University of California Press.

———. 2004. *The Wines of Bordeaux: Vintages and Tasting Notes, 1952–2003*. Berkeley: University of California Press.

Cohen, S.D., et al. 2008. "Assessing the Impact of Temperature on Grape Phenolic Metabolism." *Analytica Chimica Acta* 621 (1):57–67.

Colman, T. 2008. *Wine Politics: How Governments, Environmentalists, Mobsters, and Critics Influence the Wines We Drink*. Berkeley: University of California Press.

Colugnati, G., et al. 1997. "Comparison between Dfferent Times of Application of Slow-Release Nitrogen Fertilizers on Grapevines: Preliminary Results." *Acta Hort. (ISHS)* 448:395–402.

———. 2000. "The Importance of the 'Viticulture Model' on the Aromatic Properties of cv Sauvignon Blanc." *Acta Hort. (ISHS)* 526:415–28.

Columella. *De re rustica*. Translated by Harrison Boyd Ash. Loeb Classical Library. Cambridge, MA: Harvard University Press, 1941.

Conklin, E.G. 1922. *The Direction of Human Evolution*. Vol. 2. New ed. New York: C. Scribner's Sons.

Coombe, B.G. 1973. "The Regulation of Set and Development of the Grape Berry." *Acta Hort. (ISHS)* 34:261–74.

———. 1980. "Development of the Grape Berry 1: Effects of Time of Flowering and Competition." *Australian Journal of Agricultural Research* 31 (1):125–31.

Coombe, B.G., et al. 1987. "Solute Accumulation by Grape Pericarp Cells 5: Relationship to Berry Size and the Effects of Defoliation." *Journal of Experimental Botany* 38 (196):1789–98. doi: 10.1093/jxb/38.11.1789.

Coombe, B.G., and P.R. Monk. 1979. "Proline and Abscisic-Acid Content of the Juice of Ripe Riesling Grape Berries—Effect of Irrigation during Harvest." *American Journal of Enology and Viticulture* 30 (1):64–67.

Cordner, C.W., et al. 1978. "Effects of Crop Level on Chemical Composition and Headspace Volatiles of Lodi Zinfandel Grapes and Wines." *American Journal of Enology and Viticulture* 29 (4):247–53.

Cordner, C.W., and C.S. Ough. 1978. "Prediction of Panel Preference for Zinfandel Wine from Analytical Data: Using Difference in Crop Level to Affect Must, Wine, and Headspace Composition." *American Journal of Enology and Viticulture* 29 (4):254–57.

Cortell, J.M. 2006. "Influence of Vine Vigor and Shading in Pinot noir (Vitis vinifera L.) on the Concentration and Composition of Phenolic Compounds in Grapes and Wine." PhD diss., Oregon State University.

Coulton, G.G. 1928. *Life in the Middle Ages*. Selected, translated, and annotated by G.G. Coulton. Cambridge: Cambridge University Press.

Courchet, L. 1897. *Traité de botanique comprenant l'anatomie et la physiologie végétales et les familles naturélles a l'usage des candidats au certificat d'études physiques, chimiques et naturelles des étudiants en médecine et en pharmacie*. Vol. 1 of 2. Paris: Baillière.

Covert, J.C. 1902. "World's Wine Crop." In *Consular Reports: Commerce, Manufactures, etc.,* vol. 69. Washington, DC: Government Printing Office.

Coyle, L. P. 1982. *The World Encyclopedia of Food.* New York: Facts on File.

Creasy, G. L., and L. L. Creasy. 2009. *Grapes.* Oxfordshire, UK: CABI.

Croidieu, G. 2011. "An Inconvenient Truce: Cultural Domination and Contention after the 1855 Medoc Wine Classification Event." In *Negotiating Values in the Creative Industries: Fairs, Festivals, and Competitive Events,* edited by Brian Moeran and Jesper Strandgaard Pedersen. Cambridge: Cambridge University Press.

Cromer, A. 1995. *Uncommon Sense: The Heretical Nature of Science.* Oxford: Oxford University Press.

Dami, I., et al. 2006. "A Five-Year Study on the Effect of Cluster Thinning on Yield and Fruit Composition of 'Chambourcin' Grapevines." *Hortscience* 41 (3):586–88.

Darcy-Bertuletti, Y. 2008. *Tableau des mesures les plus courantes en usage dans le pays Beaunois.* http://www.beaune.fr/IMG/pdf/Metrologie.pdf.

Darriet, P., et al. 2000. "Identification and Quantification of Geosmin, an Earthy Odorant Contaminating Wines." *Journal of Agricultural and Food Chemistry* 48 (10):4835–38. doi: 10.1021/Jf0007683.

Debuigne, G. 1970. *Larousse des vins.* Paris: Larousse.

Deluze, A. 2010. "What Future for the Champagne Industry?" *American Association of Wine Economists Working Paper* 64:1–24.

Demossier, M. 1997. "Producing Tradition and Managing Social Changes in the French Vineyards: The Circle of Time in Burgundy." *Ethnologia Europaea* (27):47–58.

Dering, G. W. 2007. "Geomorphic Controls on Temperature Variation in the Walla Walla American Viticultural Area." Abstract of paper presented at the Cordilleran Section, 103rd Annual Meeting of the Geology Society of America, May 4–6, 2007, Bellingham, WA. https://gsa.confex.com/gsa/2007CD/finalprogram/abstract_120767.htm

Descartes, R. 2001. *Discourse on Method, Optics, Geometry, and Meteorology.* Indianapolis: Hackett Pub.

Diago, M. P., et al. 2010. "Effects of Mechanical Thinning on Fruit and Wine Composition and Sensory Attributes of Grenache and Tempranillo Varieties (Vitis vinifera L.)." *Australian Journal of Grape and Wine Research* 16 (2):314–26.

Dion, R. 1959. *Histoire de la vigne et du vin en France des origines au XXIe siècle.* Paris: Clavreuil.

Di Profio, F., et al. 2011. "Canopy Management and Enzyme Impacts on Merlot, Cabernet franc, and Cabernet Sauvignon. I. Yield and Berry Composition." *American Journal of Enology and Viticulture* 62 (2):139–51. doi: 10.5344/ajev.2010.10024.

Divirgilio, A. 2012. "Penfolds Wines Limited Edition Cabernet Sauvignon Sells for $168,000." http://www.bornrich.com/penfolds-wines-limited-edition-cabernet-sauvignon-sells-168000.html.

Dobelli, R. 2013. "The Overconfidence Effect: When You Systematically Overestimate Your Knowledge and Abilities." *Psychology Today,* June 11, 2013.

http://www.psychologytoday.com/blog/the-art-thinking-clearly/201306
/the-overconfidence-effect.

Dodds, B., and R. H. Britnell. 2008. *Agriculture and Rural Society after the Black Death: Common Themes and Regional Variations.* Studies in Regional and Local History. Hatfield: University of Hertfordshire Press.

Dokoozlian, N. K. 1990. "Light Quantity and Light Quality within Vitis vinifera L. grapevine Canopies and Their Relative Influence on Berry Growth and Composition." PhD diss., Plant Physiology, University of California, Davis.

Dokoozlian, N. K., and W. M. Kliewer. 1995. "The Light Environment within Grapevine Canopies 2: Influence of Leaf-Area Density on Fruit Zone Light Environment and Some Canopy Assessment Parameters." *American Journal of Enology and Viticulture* 46 (2):219–26.

———. 1996. "Influence of Light on Grape Berry Growth and Composition Varies during Fruit Development." *Journal of the American Society for Horticultural Science* 121 (5):869–74.

Dougherty, P. H. 2012. *The Geography of Wine: Regions, Terroir, and Techniques.* New York: Springer.

Downey, M. O., et al. 2004. "The Effect of Bunch Shading on Berry Development and Flavonoid Accumulation in Shiraz Grapes." *Australian Journal of Grape and Wine Research* 10 (1):55–73.

Dresser, M. 1992. "Vintage Point—Ridge Vineyards Turns Out Top-Quality Wines while Lying Low." *Baltimore Sun,* October 11, 1992. http://articles
.baltimoresun.com/1992–10–11/features/1992285165_1_ridge-vineyards-
monte-bello-lytton-springs.

Dry, P. R., et al. 2004. "What Is Vine Balance?" *Proceedings from the 12th Australian Wine Industry Technical Conference, Melbourne, Victoria,* 68–74.

Dry, P. R., and B. G. Coombe. 2005. *Viticulture.* Vol. 1, *Resources.* 2nd ed. Broadview: Winetitles.

Duby, G. 1992. *France in the Middle Ages, 987–1460: From Hugh Capet to Joan of Arc.* A History of France. New York: John Wiley & Sons.

———. 1998. *Rural Economy and Country Life in the Medieval West.* Philadelphia: University of Pennsylvania Press.

Duchêne, E., and C. Schneider. 2005. "Grapevine and Climatic Changes: A Glance at the Situation in Alsace." *Agronomy for Sustainable Development* 25 (1):93–99.

Due, G., et al. 1993. "Modeling Grapevine Phenology against Weather—Considerations Based on a Large Data Set." *Agricultural and Forest Meteorology* 65 (1–2):91–106.

Du Plessis, C. S. 1984. "Optimum Maturity and Quality Parameters in Grapes: A Review." *South African Journal of Enology and Viticulture* 5 (1):35–42.

Duteau, J., et al. 1981. "Influence des facteurs naturels sur la maturation du raisin, en 1979, à Pomerol et Saint-Emilion." *Conn. Vigne Vin* 15 (3): 1–27.

Duteau, J., and G. Seguin. 1973. "Caractères analytiques des sols des grands crus du Médoc." *Comptes Rendus des Séances de l'Académie d'Agriculture de France* 59 (14):1084–90.

Dyer, C. 1989. *Standards of Living in the Later Middle Ages: Social Change in England, c. 1200–1520.* Cambridge: Cambridge University Press.

Edwards, M., et al. 2009. *The Finest Wines of Champagne: A Guide to the Best Cuvées, Houses, and Growers.* Berkeley: University of California Press.

Ehrlen, J., and K. Lehtila. 2002. "How Perennial Are Perennial Plants?" *Oikos* 98 (2):308–22. doi: 10.1034/j.1600–0706.2002.980212.x.

Einstein, A. 1934. "On the Method of Theoretical Physics." *Philosophy of Science* 1 (2):163–69. doi: 10.2307/184387.

Eliot-Fissk, D.L. 1993. "Viticultural Soils of California, with Special Reference to the Napa Valley." *Journal of Wine Research* 4:67–77.

Etchebarne, F., et al. 2010. "Leaf:Fruit Ratio and Vine Water Status Effects on Grenache Noir (Vitis vinifera L.) Berry Composition: Water, Sugar, Organic Acid, and Cations." *South African Journal of Enology and Viticulture* 31 (2):106–15.

Ewart, A.J.W., et al. 1985. "The Effects of Light Pruning, Irrigation, and Improved Soil-Management on Wine Quality of the Vitis-Vinifera Cv Riesling." *Vitis* 24 (4):209–17.

Falcetti, M. 1994. "Le terroir: Qu'est-ce qu'un terroir? Pourquoi l'étudier? Pourquoi l'enseigner?" *Bull. de l'O.I.V.* 67 (2):246–75.

Fanzone, M., et al. 2011. "Phenolic Composition of Malbec Grape Skins and Seeds from Valle de Uco (Mendoza, Argentina) during Ripening: Effect of Cluster Thinning." *Journal of Agricultural and Food Chemistry* 59 (11):6120–36. doi: 10.1021/Jf200073k.

Farina, L., et al. 2005. "Terpene Compounds as Possible Precursors of 1,8-Cineole in Red Grapes and Wines." *Journal of Agricultural and Food Chemistry* 53 (5):1633–36.

Féret, E., et al. 1962. *Dictionnaire du vin.* Bordeaux: Féret et Fils.

Fischer, U., et al. 1999. "The Impact of Geographic Origin, Vintage, and Wine Estate on Sensory Properties of Vitis vinifera cv. Riesling Wines." *Food Quality and Preference* 10 (4–5):281–88. doi: 10.1016/S0950–3293(99)00008–7.

Fourcade, M. 2012. "THE VILE AND THE NOBLE: On the Relation between Natural and Social Classifications in the French Wine World." *Sociological Quarterly* 53:524–45.

Fowler, H.W. 1994. *A Dictionary of Modern English Usage.* Hertfordshire: Wordsworth Editions.

Frankfurt, H.G. 2009. *On Bullshit.* Princeton, NJ: Princeton University Press.

Franklin, B., and J. Sparks. 1840. *The Works of Benjamin Franklin: With Notes and a Life of the Author by J. Sparks.* Philadelphia: Childs & Peterson.

Franz, M. 1994. "Learning to Value Wines of Alsace." *Washington Times,* May 4, 1994.

Freedman, P.H. 2007. *Food: The History of Taste.* Berkeley: University of California Press.

Freeman, B.M. 1983. "Effects of Irrigation and Pruning of Shiraz Grapevines on Subsequent Red Wine Pigments." *American Journal of Enology and Viticulture* 34 (1):23–26.

Freeman, B.M., et al. 1979. "Interaction of Irrigation and Pruning Level on Growth and Yield of Shiraz Vines." *American Journal of Enology and Viticulture* 30 (3):218–23.

———. 1980. "Interaction of Irrigation and Pruning Level on Grape and Wine Quality of Shiraz Vines." *American Journal of Enology and Viticulture* 31 (2):124–35.

Freeman, B.M., and W.M. Kliewer. 1983. "Effect of Irrigation, Crop Level, and Potassium Fertilization on Carignane Vines II: Grape and Wine Quality." *American Journal of Enology and Viticulture* 34 (3):197–207.

Freese, P. 1999. "Vine Spacing—An International Perspective." *Proceedings, Vine Spacing Symposium, ASEV, June 29, 1999, Reno, Nevada*, 33–42.

Furetière, A. 1690. *Dictionnaire universel, contenant généralement tous les mots français tant vieux que modernes et les termes de toutes les sciences et des arts.* The Hague: A. & R. Leers. http://archive.org/stream/Dictionnaire Universel/furetiere#page/n2009/mode/2up.

Gade, D.W. 2004. "Tradition, Territory, and Terroir in French Viniculture: Cassis, France, and Appellation Contrôlée." *Annals of the Association of American Geographers* 94 (4):848–67.

Gale, A. 2004. *Jake Hancock, Geologist and Wine Researcher.* http://www.ugr .es/~mlamolda/swg/hancock.html.

Galen. *Galen: On Diseases and Symptoms.* Edited and translated by I. Johnston. Cambridge and New York: Cambridge University Press, 2006.

Gambetta, G.A., et al. 2010. "Sugar and Abscisic Acid Signaling Orthologs Are Activated at the Onset of Ripening in Grape." *Planta* 232 (1):219–34.

Gangjee, D.S. 2012. *Relocating the Law of Geographical Indications.* New York: Cambridge University Press.

Garrett, B. 2003. "Vitalism and Teleology in the Natural Philosophy of Nehemiah Grew (1641–1712)." *British Journal for the History of Science* 36 (128):63–81. doi: 10.1017/S0007087402004909.

Gately, I. 2008. *Drink: A Cultural History of Alcohol.* New York: Penguin.

Gatti, M., et al. 2012. "Effects of Cluster Thinning and Preflowering Leaf Removal on Growth and Grape Composition in cv. Sangiovese." *American Journal of Enology and Viticulture* 63 (3):325–32.

Gawel, R. 1998. "Red Wine Astringency: A Review." *Australian Journal of Grape and Wine Research* 4:74–95.

Gerbaux, V., et al. 2000. "A Study of Phenol Volatiles in Pinot noir Wines in Burgundy." *Bulletin de l'OIV* 5 (73):581–99.

Gergaud, O., and V. Ginsburgh. 2008. "Natural Endowments, Production Technologies, and the Quality of Wines in Bordeaux: Does Terroir Matter?" *Economic Journal* 118 (529):F142-F157. doi:10.1111/j.1468–0297.2008 .02146.x.

Gholami, M., et al. 1995. "Biosynthesis of Flavour Compounds in Muscat Gordo Blanco Grape Berries." *Australian Journal of Grape and Wine Research* 1 (1):19–24. doi: 10.1111/j.1755–0238.1995.tb00073.x.

Girard, B., et al. 2002. "Volatile Terpene Constituents in Maturing Gewürztraminer Grapes from British Columbia." *American Journal of Enology and Viticulture* 53 (2):99–109.

Gladstones, J. S. 1992. *Viticulture and Environment: A Study of the Effects of Environment on Grapegrowing and Wine Qualities, with Emphasis on Present and Future Areas for Growing Winegrapes in Australia.* Adelaide: Winetitles.

Glynn, M. 2003. "Distribution of Brix, Berry Weight, Seed Number, Anthocyanins, Total Skin Phenols, Skin Hydroxycinnamates, and Skin Flavonols in a Cabernet Sauvignon Cluster." Master's thesis, Viticulture and Enology, University of California, Davis.

Goheen, A. C., and J. A. Cook. 1959. "Leafroll (Red-Leaf or Rougeau) and Its Effects on Vine Growth, Fruit Quality, and Yields." *American Journal of Enology and Viticulture* 10 (4):173–81.

Goode, J. 2005. *The Science of Wine: From Vine to Glass.* Berkeley: University of California Press.

Gough, J. B. 1998. "Winecraft and Chemistry in 18th-Century France: Chaptal and the Invention of Chaptalization." *Technology and Culture* 39 (1):74–104. doi: 10.2307/3107004.

Grainger, K. 2009. *Wine Quality: Tasting and Selection.* Food Industry Briefing Series. Chichester, UK, and Ames, IA: Wiley-Blackwell.

Grant, S. 2000. "Economics of Vineyard Design: Vine and Row Spacing, Trellising." *Practical Winery & Vineyard Journal,* January/February 2000.

Gray, J. D., et al. 1994. "Assessment of Winegrape Value in the Vineyard—A Preliminary, Commercial Survey." *Wine Industry Journal* 9 (3):253–61.

———. 1997. "Assessment of Winegrape Value in the Vineyard—Survey of cv. Shiraz from South Australian Vineyards in 1992." *Australian Journal of Grape and Wine Research* 3 (3):1–8. doi: 10.1111/j.1755-0238.1997.tb00123.x.

Gray, W. B. 2014. "Do Lower Yields Mean Higher Quality?" *Palate Press—The Online Wine Magazine.* http://palatepress.com/2014/07/wine/lower-yields-mean-higher-quality/.

Grazzi-Soncini, G., and F. T. Bioletti. 1892. *Wine: Classification, Wine Tasting, Qualities, and Defects; Appendix E to the Biennial Report of the Board of State Viticultural Commissioners for 1891–92.* Sacramento: State Printing.

Greenspan, M. D. 2011. "The Indescribable Concept of Vine Balance: Reducing Vine Balance to a Number Does Little to Represent the Complex Physiological Relationship between Vine and Fruit." *Wine Business Monthly,* Sept. 2011, 66–69.

Greenspan, M. D., et al. 1994. "Developmental-Changes in the Diurnal Water-Budget of the Grape Berry Exposed to Water Deficits." *Plant Cell and Environment* 17 (7):811–20. doi: 10.1111/j.1365-3040.1994.tb00175.x.

———. 1996. "Field Evaluation of Water Transport in Grape Berries during Water Deficits." *Physiologia Plantarum* 97 (1):55–62. doi: 10.1034/j.1399-3054.1996.970109.x.

Gris, E. F., et al. 2010. "Phenology and Ripening of Vitis vinifera L. Grape Varieties in Sao Joaquim, Southern Brazil: A New South American Wine Growing Region." *Ciencia e Investigacion Agraria* 37 (2):61–75.

Gross, P. R., and N. Levitt. 1997. *Higher Superstition: The Academic Left and Its Quarrels with Science.* Baltimore: Johns Hopkins University Press.

Guild, G. 2010. "Spinoza's Conjecture." In *How Do You Think? An Exploration of Human Thought, Cognitive Biases, Neuroscience, and Quirks of the Human Brain,* January 22, 2010. http://geraldguild.com/blog/2010/01/22/spinozas-conjecture/.

Guy, K. M. 2003. *When Champagne Became French: Wine and the Making of a National Identity.* Johns Hopkins University Studies in Historical and Political Science. Baltimore: Johns Hopkins University Press.

Guyot, C., and P. Dupraz. 2004. "Déguster les baies pour suivre la maturité des raisins." *Rev. Suisse Vitic. Arboric. Hortic.* 36 (4):231–34.

Guyot, J. 2010. *Culture de la vigne et vinification* (1861). Whitefish: Kessinger Publishing.

Haeger, J. W. 2004. *North American Pinot Noir.* Berkeley: University of California Press.

Hailman, J. R. 2006. *Thomas Jefferson on Wine.* Jackson: University Press of Mississippi.

Hake, S. J., et al. 1996. *Cotton Production Manual.* Oakland: University of California, Division of Agriculture and Natural Resources.

Hale, C. R., and M. S. Buttrose. 1974. "Effect of Temperature on Ontogeny of Berries of Vitis-Vinifera L Cv Cabernet-Sauvignon." *Journal of the American Society for Horticultural Science* 99 (5):390–94.

Hales, S. 1727. *Vegetable Staticks: Or, An Account of Some Statical Experiments On The Sap in Vegetables: Being an Essay Towards a Natural History of Vegetation. Also, a Specimen of An Attempt to Analyse the Air, By a Great Variety of Chymio-Statical Experiments; Which Were Read at Several Meetings Before the Royal Society.* Vol. 1. London: W. and J. Innys; T. Woodward.

Hall, A. D., and C. G. T. Morison. 1906. "On the Function of Silica in the Nutrition of Cereals: Part I." *Proceedings of the Royal Society of London, Series B, Containing Papers of a Biological Character* 77 (520):455–77.

Halliday, J. 2006. *Wine Atlas of Australia.* Berkeley: University of California Press.

Halliday, J., and H. Johnson. 2007. *The Art and Science of Wine.* Richmond Hill, Ont.: Firefly Books.

Hamman, R. A., and I. E. Dami. 1999. "Evaluation of 35 Wine Grape Cultivars and 'Chardonnay' on 4 Rootstocks Grown in Western Colorado." Fort Collins: Colorado State University Agricultural Experiment Station.

Hancock, D. 2009. *Oceans of Wine: Madeira and the Emergence of American Trade and Taste.* New Haven, CT: Yale University Press.

Hanni, T. 2013. *Why You Like the Wines You Like: Changing the Way the World Thinks about Wine.* The New Wine Fundamentals, vol. 1. Napa, CA: HanniCo.

Hanson, V. D. 1995. *The Other Greeks: The Family Farm and the Agrarian Roots of Western Civilization.* New York: Simon and Schuster.

Hardie, W. J., et al. 1996. "Morphology, Anatomy, and Development of the Pericarp after Anthesis in Grape, Vitis vinifera L." *Australian Journal of Grape and Wine Research* 2 (2):97–142. doi: 10.1111/j.1755-0238.1996.tb00101.x.

Haynes, S. J. 1999. "Geology and Wine 1: Concept of Terroir and the Role of Geology." *Geoscience Canada* 26 (4):190–94.

Heap, A. J., and E. I. Newman. 1980. "The Influence of Vesicular-Arbuscular Mycorrhizas on Phosphorus Transfer between Plants." *New Phytologist* 85 (2):173–79. doi: 10.1111/j.1469–8137.1980.tb04458.x.

Heazlewood, J. E., et al. 2006. "Pruning Effects on Pinot Noir Vines in Tasmania (Australia)." *Vitis* 45 (4):165–71.

Hedberg, P. R., and J. Raison. 1982. "The Effect of Vine Spacing and Trellising on Yield and Fruit Quality of Shiraz Grapevines." *American Journal of Enology and Viticulture* 33 (1):20–30.

Henick-Kling, T., et al. 2000. "Brettanomyces in Wine." Paper presented at the 5th International Symposium on Cool Climate Viticulture and Oenology, January 16–20, Melbourne, Australia.

Henzell, T. 2007. *Australian Agriculture: Its History and Challenges.* Collingwood, Vic.: CSIRO Publishing.

Hepner, Y., et al. 1985. "Effect of Drip Irrigation Schedules on Growth, Yield, Must Composition, and Wine Quality of Cabernet Sauvignon." *American Journal of Enology and Viticulture* 36 (1):77–85.

Herrera, E. 2004. "Pruning the Home Orchard." Cooperative Extension Service Guide H-237, College of Agriculture and Home Economics, New Mexico State University. http://aces.nmsu.edu/pubs/_h/H327.pdf.

Hershey, D. 2003. "Misconceptions about Helmont's Willow Experiment." *Plant Science Bulletin* (49):63–81.

Heymann, H., et al. 2013. "Effects of Extended Grape Ripening with or without Must and Wine Alcohol Manipulations on Cabernet Sauvignon Wine Sensory Characteristics." *South African Journal of Enology and Viticulture* 34 (1):86–99.

Hill, C. O. 2011. "On Vine Spacing: Dear Andrew." *Grape Press* 27 (1): 7. http://www.virginiavineyardsassociation.com/wp-content/uploads/2011/06/GrapePress-April-2011.pdf.

Hillel, D. 1987. "On the Tortuous Path of Research." *Soil Science* 143 (4): 304–5.

Hinton, R. J., and United States Department of Agriculture. 1890. *Irrigation in the United States.* U.S. Congress Senate Select Committee on Irrigation and Reclamation of Arid Lands Report. Washington, DC: Government Printing Office.

Hirsch, J. 2005. "Winemakers, Growers Fight over 'Hang Time.'" *Los Angeles Times,* January 10, 2005. http://articles.latimes.com/2005/jan/10/business/fi-hangtime10.

Hodson, A. 2011. "Vine Spacing: An Opinion from the VVA Annual Technical Meeting 2011." *Grape Press* 27(1): 1. http://www.virginiavineyardsassociation.com/wp-content/uploads/2011/06/GrapePress-April-2011.pdf.

Holt, H. E., et al. 2008. "Relationships between Berry Size, Berry Phenolic Composition, and Wine Quality Scores for Cabernet Sauvignon (Vitis vinifera L.) from Different Pruning Treatments and Different Vintages." *Australian Journal of Grape and Wine Research* 14(3):191–202. doi: 10.1111/j.1755–0238.2008.00019.x.

Holton, G. J. 1993. *Science and Anti-science.* Cambridge, MA: Harvard University Press.

Holzapfel, B., et al. 1999. "Ripening Grapes to Specification: Effect of Yield on Colour Development of Shiraz Grapes in the Riverina." *Australian Grapegrower and Winemaker* 428:24–28.

Hooke, H. 1991. "Great Grapes Grow in Gravel." *Sydney Morning Herald,* May 7, 1991.

Howell, G. S. 2001. "Sustainable Grape Productivity and the Growth-Yield Relationship: A Review." *American Journal of Enology and Viticulture* 52 (3):165–74.

Hughson, A. L., and R. A. Boakes. 2002. "The Knowing Nose: The Role of Knowledge in Wine Expertise." *Food Quality and Preference* 13 (7–8): 463–72.

Hummell, A. K., and D. C. Ferree. 1998. "Interaction of Crop Level and Fruit Cluster Exposure on 'Seyval blanc' Fruit Composition." *Journal of the American Society for Horticultural Science* 123 (5):755–61.

Hunter, J. J. 1998. "Plant Spacing Implications for Grafted Grapevine II: Soil Water, Plant Water Relations, Canopy Physiology, Vegetative and Reproductive Characteristics, Grape Composition, Wine Quality, and Labour Requirements." *South African Journal of Enology and Viticulture* 19 (2):35–51.

Hunter, J. J., and A. C. De La Harpe. 1987. "The Effect of Rootstock Cultivar and Bud Load on the Colour of Vitis vinifera L. cv. Muscat noir (Red Muscadel) Grapes." *South African Journal of Enology and Viticulture* 8:1–5.

Husenbeth, F. C. 1834. *A Guide for the Wine Cellar, or, a Practical Treatise on the Cultivation of the Vine, and the Management of the Different Wines Consumed in This Country.* London: Effingham Wilson, Royal Exchange.

Husmann, G. 1888. *Grape Culture and Wine-Making in California: A Practical Manual for the Grape-Grower and Wine-Maker.* San Francisco: Payot, Upham.

Iland, P. G., and B. G. Coombe. 1988. "Malate, Tartrate, Potassium, and Sodium in Flesh and Skin of Shiraz Grapes during Ripening: Concentration and Compartmentation." *American Journal of Enology and Viticulture* 39 (1):71–76.

Intrieri, C. 1987. "Experiences on the Effect of Vine Spacing and Trellis-Training System on Canopy Micro-Climate, Vine Performance, and Grape Quality." *Acta Hort. (ISHS)* 206:69–88.

Intrieri, C., and I. Filippetti. 2000. "Planting Density and Physiological Balance: Comparing Approaches to European Viticulture in the 21st Century." *Proceedings of the American Society of Enology and Viticulture 50th Anniversary Annual Meeting, June 19–23, 2000, Seattle, Washington,* 296–308.

Intrigliolo, D. S., and J. R. Castel. 2011. "Interactive Effects of Deficit Irrigation and Shoot and Cluster Thinning on Grapevine cv. Tempranillo: Water Relations, Vine Performance, and Berry and Wine Composition." *Irrigation Science* 29 (6):443–54.

Jackson, D. I., and P. B. Lombard. 1993. "Environmental and Management Practices Affecting Grape Composition and Wine Quality—a Review." *American Journal of Enology and Viticulture* 44 (4):409–30.

Jackson, R. S. 2000. *Wine Science: Principles, Practice, Perception.* San Diego: Academic Press.

———. 2008. *Wine Science: Principles and Applications.* San Diego, CA: Elsevier Science.

Jacob, H. E., and A. J. Winkler. 1950. *Grape Growing in California.* Berkeley: College of Agriculture, University of California.

Jekel, B., and B. Prats. 1983. "How Important Is 'Terroir' in a Fine Wine?" *Decanter* 8 (12):22–27.

Jensen, E. M. 2006. Review of *The Chief Purpose of Universities: Academic Discourse and the Diversity of Ideas,* by William M. Bowen and Michael Schwartz. *Cleveland State Law Review* 54:393–404.

Johnson, H. 1983. *Hugh Johnson's Modern Encyclopedia of Wine.* New York: Simon and Schuster.

———. 1989. *Vintage: The Story of Wine.* New York: Simon and Schuster.

———. 1999. *The Peter Allan Sichel Memorial Lecture.* http://www.wsetglobal .com/documents/hughj99.pdf.

———. 2004. *Story of Wine.* London: Octopus Books.

Johnstone, R. S., et al. 1995. "The Composition of Shiraz Grape Berries—Implications for Wine." *Proceedings from the 4th Australian Wine Industry Technical Conference, Adelaide, South Australia,* 105–8.

Jones, G. V., et al. 2012. "Climate, Grapes, and Wine: Structure and Suitability in a Variable and Changing Climate." In *The Geography of Wine,* edited by P. H. Dougherty. New York: Springer.

Jones, G. V., and R. E. Davis. 2000. "Climate Influences on Grapevine Phenology, Grape Composition, and Wine Production and Quality for Bordeaux, France." *American Journal of Enology and Viticulture* 51 (3):249–61.

Jones, P. 1988. *The Peasantry in the French Revolution.* Cambridge and New York: Cambridge University Press.

Joseph, R. 1996. *The Book of Wine.* New York: Smithmark Pub.

Josling, T. 2006. "The War on Terroir: Geographical Indications as a Transatlantic Trade Conflict." *Journal of Agricultural Economics* 57 (3):337–63. doi: 10.1111/j.1477-9552.2006.00075.x.

Jullien, A. 1866. *Topographie de tous les vignobles connus, contenant leur position géographique.* Paris: Imprimerie et Librairie d'agriculture.

Kalua, C. M., and P. K. Boss. 2010. "Comparison of Major Volatile Compounds from Riesling and Cabernet Sauvignon Grapes (Vitis vinifera L.) from Fruitset to Harvest." *Australian Journal of Grape and Wine Research* 16 (2):337–48.

Kasimatis, A. N., et al. 1977. "Relationship of Soluble Solids and Berry Weight to Airstream Grades of Natural Thompson Seedless Raisins." *American Journal of Enology and Viticulture* 28 (1):8–15.

———. 1985. "Conversion of Cane-Pruned Cabernet Sauvignon Vines to Bilateral Cordon Training and a Comparison of Cane and Spur Pruning." *American Journal of Enology and Viticulture* 36 (3):240–44.

Keller, M. 2005. "Deficit Irrigation and Vine Mineral Nutrition." *American Journal of Enology and Viticulture* 56 (3):267–83.

Keller, M., et al. 1998. "Interaction of Nitrogen Availability during Bloom and Light Intensity during Veraison I: Effects on Grapevine Growth, Fruit

Development, and Ripening." *American Journal of Enology and Viticulture* 49 (3):333–40.

———. 2004. "Crop Load Management in Concord Grapes Using Different Pruning Techniques." *American Journal of Enology and Viticulture* 55 (1):35–50.

———. 2005. "Cluster Thinning Effects on Three Deficit-Irrigated Vitis vinifera Cultivars." *American Journal of Enology and Viticulture* 56 (2):91–103.

———. 2008. "Interactive Effects of Deficit Irrigation and Crop Load on Cabernet Sauvignon in an Arid Climate." *American Journal of Enology and Viticulture* 59 (3):221–34.

———. 2010. "Spring Temperatures Alter Reproductive Development in Grapevines." *Australian Journal of Grape and Wine Research* 16 (3):445–54. doi: 10.1111/j.1755-0238.2010.00105.x.

Keller, M., and G. Hrazdina. 1998. "Interaction of Nitrogen Availability during Bloom and Light Intensity during Veraison II: Effects on Anthocyanin and Phenolic Development during Grape Ripening." *American Journal of Enology and Viticulture* 49 (3):341–49.

Kennedy, J. 2002. "Understanding Grape Berry Development." *Practical Winery & Vineyard Journal,* July/August 2002. http://www.practicalwinery.com/julyaugust02/julaugo2p14.htm.

Kevany, S. 2008. "New Champagne Areas Defined." March 14, 2008. http://www.decanter.com/news/wine-news/485939/new-champagne-areas-defined.

Klebs, G. 1910. "Alterations of the Development and Forms of Plants as a Result of Environment." *Nature* 83:414.

Kliewer, W. M. 1970. "Effect of Time and Severity of Defoliation on Growth and Composition of 'Thompson Seedless' Grapes." *American Journal of Enology and Viticulture* 21 (1):37–47.

———. 1971. "Effect of Nitrogen on Growth and Composition of Fruits from Thompson Seedless Grapevines." *Journal of the American Society for Horticultural Science* 96 (6):816.

———. 1991. "Vineyard Canopy Management Practices for Improving Grapevine." http://www.avf.org/article.html?id=3607.

Kliewer, W. M., et al. 1983. "Effect of Irrigation, Crop Level, and Potassium Fertilization on Carignane Vines I: Degree of Water Stress and Effect on Growth and Yield." *American Journal of Enology and Viticulture* 34 (3):186–96.

———. 2000. "Trellis and Vine Spacing Effects on Growth, Canopy Microclimate, Yield, and Fruit Composition of Cabernet Sauvignon." *Acta Hort. (ISHS)* 526:21–32.

Kliewer, W. M., and A. J. Antcliff. 1970. "Influence of Defoliation, Leaf Darkening, and Cluster Shading on the Growth and Composition of Sultana Grapes." *American Journal of Enology and Viticulture* 21 (1):26–36.

Kliewer, W. M., and J. A. Cook. 1974. "Arginine Levels in Grape Canes and Fruits as Indicators of Nitrogen Status of Vineyards." *American Journal of Enology and Viticulture* 25 (2):111–18.

Kliewer, W. M., and N. K. Dokoozlian. 2005. "Leaf Area/Crop Weight Ratios of Grapevines: Influence on Fruit Composition and Wine Quality." *American Journal of Enology and Viticulture* 56 (2):170–81.

Kliewer, W. M., and R. D. Fuller. 1973. "Effect of Time and Severity of Defoliation on Growth of Roots, Trunk, and Shoots of 'Thompson Seedless' Grapevines." *American Journal of Enology and Viticulture* 24 (2):59–64.

Kliewer, W. M., and L. A. Lider. 1970. "Effects of Day Temperature and Light Intensity on Growth and Composition of Vitis-Vinifera L Fruits." *Journal of the American Society for Horticultural Science* 95 (6):766-69.

Kliewer, W. M., and C. S. Ough. 1970. "Effect of Leaf Area and Crop Level on the Concentration of Amino Acids and Total Nitrogen in Thompson Seedless Grapes." *Vitis* 9:196–206.

Kliewer, W. M., and H. B. Schultz. 1973. "Effect of Sprinkler Cooling of Grapevines on Fruit Growth and Composition." *American Journal of Enology and Viticulture* 24 (1):17–26.

Kliewer, W. M., and R. J. Weaver. 1971. "Effect of Crop Level and Leaf Area on Growth, Composition, and Coloration of 'Tokay' Grapes." *American Journal of Enology and Viticulture* 22 (3):172–77.

Knight, T. A. 1837. "Upon the Supposed Absorbent Powers of the Cellular Points, or Spongioles, of the Roots of Trees, and Other Plants." *London and Edinburgh Philosophical Magazine and Journal of Science* 10:488.

Knops, J. M., et al. 2007. "Negative Correlation Does Not Imply a Tradeoff between Growth and Reproduction in California Oaks." *Proc. Natl. Acad. Sci. U.S.A.* 104 (43):16982–85. doi: 10.1073/pnas.0704251104.

Koch, A., et al. 2010. "2-Methoxy-3-Isobutylpyrazine in Grape Berries and Its Dependence on Genotype." *Phytochemistry* 71 (17–18):2190–98.

———. 2012. "Fruit Ripening in Vitis vinifera: Light Intensity before and Not during Ripening Determines the Concentration of 2-Methoxy-3-Isobutylpyrazine in Cabernet Sauvignon Berries." *Physiologia Plantarum* 145 (2):275–85.

Kozlowski, T. T., and S. G. Pallardy. 1997. *Physiology of Woody Plants*. 2nd ed. San Diego: Academic Press.

Kramer, M. 1990. *Making Sense of Burgundy*. New York: W. Morrow.

———. 2004. *Matt Kramer's New California Wine*. Philadelphia: Running Press.

Kramer, P. J. 1956. "The Role of Physiology in Forestry." *Forestry Chronicle* 32 (3):297–308.

———. 1983. *Water Relations of Plants*. San Diego: Academic Press.

Krasnow, M., et al. 2008. "Evidence for Substantial Maintenance of Membrane Integrity and Cell Viability in Normally Developing Grape (Vitis vinifera L.) Berries throughout Development." *Journal of Experimental Botany* 59 (4):849–59.

Krupinska, K., et al. 2003. "Genetic, Metabolic, and Environmental Factors Associated with Aging in Plants." In *Aging of Organisms*, edited by Heinz D. Osiewacz. Boston: Kluwer Academic Publishers.

Kunzig, R. 1999. "The Chemistry of . . . Wine Making: Is Dirt Destiny? Do the French Know What They're Doing?," *Discover*, April 1, 1999.

Labbé, E. 1938. "La participation régionales." *Rapport Général* 8:317.

Lachiver, M. 1988. *Vins vignes et vignerons, histoire du vignoble francais*. Paris: Fayard.

Lafforgue, G., and G. Chappaz. 1947. *Le vignoble girondin*. Paris: Louis Larmat.

Langewiesche, W. 2000. "The Million Dollar Nose." *Atlantic Monthly* 286 (6):42–70. http://www.theatlantic.com/past/docs/issues/2000/12/langewiesche.htm.

Laville, P. 1990. "The 'Terroir', an Indispensable Concept to the Elaboration and Protection of Appellation of Origin as to the Management of Vineyards: The Case of France." *Bulletin de l'OIV* 63 (709–10):218–40.

Lawther, J., et al. 2010. *The Finest Wines of Bordeaux: A Regional Guide to the Best Châteaux and Their Wines*. Berkeley: University of California Press.

Lee, H. 1887. *The Vegetable Lamb of Tartary: A Curious Fable of the Cotton Plant, to Which Is Added a Sketch of the History of Cotton and the Cotton Trade* London: S. Low, Marston, Searle, & Rivington.

Leicester, H.M., and H.S. Klickstein. 1963. *A Source Book in Chemistry, 1400–1900*. Source Books in the History of the Sciences. Cambridge, MA: Harvard University Press.

Lichine, A. 1967. *Alexis Lichine's Encyclopedia of Wines & Spirits*. New York: Alfred A. Knopf.

Lider, L.A., et al. 1973. "Effect of Pruning Severity and Rootstock on Growth and Yield of 2 Grafted, Cane-Pruned Wine Grape Cultivars." *Journal of the American Society for Horticultural Science* 98 (1):8–11.

———. 1975. "Effect of Pruning Severity on the Growth and Fruit Production of 'Thompson Seedless' Grapevines." *American Journal of Enology and Viticulture* 26 (4):175–78.

Lieber, F., et al. 1832. *Encyclopedia Americana: A Popular Dictionary of Arts, Sciences, Literature, History, Politics, and Biography, Brought down to the Present Time; Including a Copious Collection of Original Articles in American Biography*. Vol. 12. Philadelphia: Carey and Lea.

Liebig, J. 1847. *Chemistry in Its Application to Agriculture and Physiology*. Edited by L.P.B. Playfair. Philadelphia: T.B. Peterson.

Lindley, D.V. 2006. "Analysis of a Wine Tasting." *Journal of Wine Economics* 1 (1):33–41.

Lines-Kelly, R. 2004. *Soil: Our Common Ground—a Humanities Perspective*. The Australian Society of Soil Science, Inc. and the New Zealand Society of Soil Science. http://www.regional.org.au/au/asssi/supersoil2004/keynote/lineskelly.htm.

Livingston, B.E., and G.J. Livingston. 1913. "Temperature Coefficients in Plant Geography and Climatology." *Botanical Gazette* 56:349–75.

Logan, W.B. 2007. *Dirt: The Ecstatic Skin of the Earth*. New York: W.W. Norton.

Loinger, C., and B. Safron. 1971. "Interdependence of Vine Yield, Grape Maturation, and Wine Quality." *Annales de Tech. Agricole* 20:225–40.

Loubère, L.A. 1978. *The Red and the White: The History of Wine in France and Italy in the Nineteenth Century*. Albany: State University of New York Press.

Loudon, J.C. 1826. *An Encyclopædia of Agriculture: Comprising the Theory and Practice of the Valuation, Transfer, Laying Out, Improvement, and*

Management of Landed Property; and the Cultivation and Economy of the Animal and Vegetable Productions of Agriculture, Including All the Latest Improvements; a General History of Agriculture in All Countries; and a Statistical View of Its Present State, with Suggestions for Its Future Progress in the British Isles. London: Longman, Hurst, Rees, Orme, Brown, and Green.

———. 1886. *An Encyclopædia of Agriculture.* 8th ed. London: Longman, Brown, Green, and Longmans.

Loughrin, J.H., and M.J. Kasperbauer. 2001. "Light Reflected from Colored Mulches Affects Aroma and Phenol Content of Sweet Basil (Ocimum basilicum L.) Leaves." *Journal of Agricultural and Food Chemistry* 49 (3):1331–35. doi: 10.1021/Jf0012648.

Lynn, C.D., and F.L. Jensen. 1966. "Thinning Effects of Bloomtime Gibberellin Sprays on Thompson Seedless Table Grapes." *American Journal of Enology and Viticulture* 17 (4):283–89.

Maccarrone, G., et al. 1996. "Assessment of Source-Sink Relationships with Simple Indices in Grapevine." *Acta Hort. (ISIIS)* 427:177–86.

MacGovern, P.E., et al. 2000. *The Origins and Ancient History of Wine.* Newark: Gordon and Breach Publishers.

Mackie, H.G. 1912. "Cotton Growing in Argentina." *Great Britain Board of Trade Journal* 78:577–79.

MacNeil, K. 1998. "A Stag's Leap Wine Is Like Falling into a Pot of Blackberry Jam." *Los Angeles Times*, July 8, 1998.

———. 2001. *The Wine Bible.* New York: Workman Publishing.

Magiels, G. 2010. *From Sunlight to Insight: Jan IngenHousz, the Discovery of Photosynthesis and Science in the Light of Ecology.* Brussels: VUBPRESS.

Main, G.L., and J.R. Morris. 2004. "Leaf-Removal Effects on Cynthiana Yield, Juice Composition, and Wine Composition." *American Journal of Enology and Viticulture* 55 (2):147–52.

Maltman, A. 2008. "The Role of Vineyard Geology in Wine Typicity." *Journal of Wine Research* 19 (1):1–17.

Marais, J. 1983. "Terpenes in the Aroma of Grapes and Wines: A Review." *South African Journal of Enology and Viticulture* 4 (2):49–58.

Mariotte, E. 1740. *De la végétation de plantes.* In *Oeuvres de m. Mariotte, de l'Académie royale des sciences,* new ed., vol. 1. La Haye, Batavia: Jean Neaulme.

Matthews, M.A., et al. 1987. "Phenologic and Growth-Responses to Early and Late Season Water Deficits in Cabernet Franc." *Vitis* 26 (3):147–60.

———. 1990. "Dependence of Wine Sensory Attributes on Vine Water Status." *Journal of the Science of Food and Agriculture* 51 (3):321–35. doi: 10.1002/jsfa.2740510305.

Matthews, M.A., and M.M. Anderson. 1988. "Fruit Ripening in Vitis vinifera L.: Responses to Seasonal Water Deficits." *American Journal of Enology and Viticulture* 39 (4):313–20.

———. 1989. "Reproductive Development in Grape (Vitis vinifera L.): Responses to Seasonal Water Deficits." *American Journal of Enology and Viticulture* 40 (1):52–60.

Matthews, M. A., and J. S. Boyer. 1984. "Acclimation of Photosynthesis to Low Leaf Water Potentials." *Plant Physiology* 74 (1):161–66.

Matthews, M. A., and V. Nuzzo. 2007. "Berry Size and Yield Paradigms on Grapes and Wine Quality." *Acta Hort. (ISHS)* 754:423–36.

Maume, L., and J. Dulac. 1945. "Potassic Deficiency in the Vine Disclosed by Chemical Analysis of the Leaf before the Appearance of 'Brunissure'" [in French]. *C. R. Hebd. Seances Acad. Sci.* 221:116–18.

May, P., et al. 1973. "Effect of Various Combinations of Trellis, Pruning, and Rootstock on Vigorous Sultana Vines." *Vitis* 12:192–206.

Mayr, E. 2004. *What Makes Biology Unique? Considerations on the Autonomy of a Scientific Discipline.* Cambridge: Cambridge University Press.

Mazza, G., et al. 1999. "Anthocyanins, Phenolics, and Color of Cabernet Franc, Merlot, and Pinot noir Wines from British Columbia." *Journal of Agricultural and Food Chemistry* 47 (10):4009–17.

McCoy, E. 2005. *The Emperor of Wine: The Rise of Robert M. Parker Jr. and the Reign of American Taste.* New York: HarperCollins Publishers.

McGovern, P. E. 2003. *Ancient Wine: The Search for the Origins of Viniculture.* Princeton, NJ: Princeton University Press.

McIntyre, G. N., et al. 1982. "The Chronological Classification of Grapevine Phenology." *American Journal of Enology and Viticulture* 33 (2):80–85.

———. 1987. "Some Limitations of the Degree Day System as Used in Viticulture in California." *American Journal of Enology and Viticulture* 38 (2):128–32.

Medawar, P. B. 1984. *The Limits of Science.* 1st ed. New York: Harper & Row.

Meloni, G., and J. Swinnen. 2012. "The Political Economy of European Wine Regulations." *LICOS Discussion Paper Series,* October 17, 2012. https://www.econ.kuleuven.be/licos/publications/dp/dp320.pdf.

Mengel, K., and E. A. Kirkby. 2001. *Principles of Plant Nutrition.* Boston: Kluwer Academic Publishers.

Michel, J. B., et al. 2011. "Quantitative Analysis of Culture Using Millions of Digitized Books." *Science* 331 (6014):176–82. doi: 10.1126/science.1199644.

Miller, D. P., and G. S. Howell. 1998. "Influence of Vine Capacity and Crop Load on Canopy Development, Morphology, and Dry Matter Partitioning in Concord Grapevines." *American Journal of Enology and Viticulture* 49 (2):183–90.

Miller, E. V. 1946. "The Physiology of Citrus Fruits in Storage." *Proceedings of the Florida State Horticultural Society* 58:128–33.

Mohr, F. 1867. *The Grape Vine: A Practically Scientific Treatise on Its Management.* New York: O. Judd.

Monroe, M. 2007. "Buzz-Worthy Sparkling Wines." *Forbes,* March 12, 2007. http://www.forbes.com/2007/04/12/fizzy-wine-champagne-forbeslife-cx_mm_0413sparklers.html.

Moran, W. 1993. "The Wine Appellation as Territory in France and California." *Annals of the Association of American Geographers* 83 (4):694–717. doi: 10.1111/j.1467-8306.1993.tb01961.x.

———. 2001. "Terroir—the Human Factor." *Wine Industry Journal* 2:32–51.

Morford, S. L., et al. 2011. "Increased Forest Ecosystem Carbon and Nitrogen Storage from Nitrogen Rich Bedrock." *Nature* 477 (7362):78-81. doi: 10.1038/Nature10415.

Mori, K., et al. 2007. "Loss of Anthocyanins in Red-Wine Grape under High Temperature." *Journal of Experimental Botany* 58 (8):1935–45.

Morlat, R., and F. Bodin. 2006. "Characterization of Viticultural Terroirs Using a Simple Field Model Based on Oil Depth—II. Validation of the Grape Yield and Berry Quality in the Anjou Vineyard (France)." *Plant and Soil* 281 (1–2):55-69.

Morley, H. 1852. *Palissy the Potter: The Life of Bernard Palissy.* Vol. 2. Boston: Ticknor, Reed and Fields.

Morris, R. 2011. "Is the Reign of Terroir Over? The Pinot Cubed Experiment." *Isante Magazine,* January 28, 2011. http://www.isantemagazine.com/prod/blog/reign-terroir-over-pinot-cubed-experiment.

Morton, L. 2011. "Vine Spacing Response." *Grape Press* 27 (1): 6. http://www.virginiavineyardsassociation.com/wp-content/uploads/2011/06/GrapePress-April-2011.pdf.

Mott, V. 1806. "An Experimental Inquiry into the Chemical and Medical Properties of the Statice Limonium of Linnaeus." MD diss., Columbia College. https://archive.org/stream/2564038R.nlm.nih.gov/2564038R#page/n5/mode/2up.

Mueller, R. A. E., and D. A. Sumner. 2006. "Clusters of Grapes and Wine." Paper presented at the Third International Wine Business Research Conference, July 6–8, Montpellier, France. www.agmrc.org/media/cms/Wine_Clusters2_84CD1EE476398.pdf.

Muller-Thurgau, H. 1898. "Abhängigkeit der Ausbildung der Traubenbeeren und einiger anderer Früchte von der Entwicklung der Samen." *Landw. Jahrb. Schweiz* 12:135–205.

Munné-Bosch, S. 2008. "Do Perennials Really Senesce?" *Trends in Plant Science* 13 (5):216–20. doi: 10.1016/j.tplants.2008.02.002.

Munson, T. V. 1909. *Foundations of American Grape Culture.* New York: Orange Judd.

Murisier, F., et al. 2005. "Experimental Trials on Training and Pruning Systems on Merlot Vines in Ticino, Switzerland: Agronomic Response and Wine Quality" [in French]. *Revue Suisse de Viticulture, Arboriculture et Horticulture* 37 (4):209–14.

Murisier, F., and M. Ferretti. 1996. "Influence of Vine Spacing in the Row on Grape Production and Quality: A Trial with Merlot Grape in the Tessin" [in French]. *Revue Suisse de Viticulture, Arboriculture et Horticulture* 28 (5):293–300.

Murisier, F., and V. Zufferey. 2003. "Influence of Plantation Density on the Agronomic Behaviour of Grapevine and on the Quality of Wines: Study on Chasselas 1, Agronomic Results." *Revue Suisse de Viticulture, Arboriculture et Horticulture* 35 (6):341–48.

Myers, B. R. 2002. *A Reader's Manifesto: An Attack on the Growing Pretentiousness in American Literary Prose.* Brooklyn: Melville House.

Myles, S., et al. 2011. "Genetic Structure and Domestication History of the Grape." *Proceedings of the National Academy of Sciences of the United States of America* 108 (9):3530–35.

Nabokov, V. V. 1981. *Lectures on Russian Literature*. Edited by F. Bowers. New York and London: Harcourt Brace Jovanovich/Bruccoli Clark.

Naor, A., et al. 2002. "Shoot and Cluster Thinning Influence Vegetative Growth, Fruit Yield, and Wine Quality of 'Sauvignon blanc' Grapevines." *Journal of the American Society for Horticultural Science* 127 (4):628–34.

Nardozza, S., et al. 2010. "Variation in Carbon Content and Size in Developing Fruit of Actinidia deliciosa Genotypes." *Functional Plant Biology* 37 (6):545–54. doi: 10.1071/Fp09301.

Negrul, A. M. 1938. "Evolution of Cultivated Forms of Grapes." *Comptes Rendus de l'Académie des Sciences de l'Urss* 18:585–88.

———. 1946. *Origin and Classification of Cultured Grape*. Edited by A. Baranov et al. Vol. 1 of *The Ampelography of the USSR*. Moscow: Pischepromizdat.

Nemani, R. R., et al. 2001. "Asymmetric Warming over Coastal California and Its Impact on the Premium Wine Industry." *Climate Research* 19 (1):25–34. doi: 10.3354/Cro19025.

Neuhauser, S., et al. 2009. "Sorosphaera viticola, a Plasmodiophorid Parasite of Grapevine." *Phytopathologia Mediterranea* 48 (1):136–39.

New York State Agricultural Experiment Station, and New York Wine Grape Foundation. 1990. *Grape Research News*. Geneva, NY: NYS Agricultural Experiment Station.

Nicolson, A. B., and C. Hanley. 1953. "Indices of Physiological Maturity: Derivation and Interrelationships." *Child Development* 24 (1):3–38.

Nii, N., and B. G. Coombe. 1983. "Structure and Development of the Berry and Pedicel of the Grape Vitis Vinifera L." *Acta Hort. (ISHS)* 139:129–40.

Nowak, B., and B. Wichman. 2005. *The Everything Wine Book: From Chardonnay to Zinfandel—All You Need to Make the Perfect Choice*. Avon: F+W Media.

Nuland, S. B. 2001. *The Mysteries Within: A Surgeon Explores Myth, Medicine, and the Human Body*. New York: Simon & Schuster.

Nuzzo, V., and M. A. Matthews. 2006. "Response of Fruit Growth and Ripening to Crop Level in Dry-Farmed Cabernet Sauvignon on Four Rootstocks." *American Journal of Enology and Viticulture* 57 (3):314–24.

Obeso, J. R. 2002. "The Costs of Reproduction in Plants." *New Phytologist* 155 (3):321–48.

Oczkowski, E. 2006. "Modeling Winegrape Prices in Disequilibrium." *Agricultural Economics* 34 (1):97–107. doi: 10.1111/j.1574-0862.2006.00107.x.

O'Daniel, S. B., et al. 2012. "Effects of Balanced Pruning Severity on Traminette (Vitis spp.) in a Warm Climate." *American Journal of Enology and Viticulture* 63 (2):284–90. doi: 10.5344/ajev.2012.11056.

Olken, C. 2012. "Five Myths about California Wine Die Hard." In *Connoisseurs' Guide to California Wine*, October 29 2012. http://www.cgcw.com/databaseshowitem.aspx?id=79500.

Olle, D., et al. 2011. "Effect of Pre- and Post-veraison Water Deficit on Proanthocyanidin and Anthocyanin Accumulation during Shiraz Berry Development." *Australian Journal of Grape and Wine Research* 17 (1):90–100.

Olmo, H. P. 1979. "Vineyard in the Year 2000: Technical Pressures." *Acta Hort. (ISHS)* 104:11–20.

Orloff, S. B., and D. H. Putnam. 2007. "Forage Quality and Testing." In *Irrigated Alfalfa Management in Mediterranean and Desert zones,* edited by C. G. Summers and D. H. Putnam. Oakland: University of California Agriculture and Natural Resources Publication.

Ortega-Regules, A., et al. 2006. "Anthocyanin Fingerprint of Grapes: Environmental and Genetic Variations." *Journal of the Science of Food and Agriculture* 86 (10):1460–67. doi: 10.1002/Jsfa.2511.

Osborne, L. 2004. *The Accidental Connoisseur: An Irreverent Journey through the Wine World.* New York: North Point Press.

Ough, C. S., and T. H. Lee. 1981. "Effect of Vineyard Nitrogen-Fertilization Level on the Formation of Some Fermentation Esters." *American Journal of Enology and Viticulture* 32 (2):125–27.

Ough, C. S., and Richard Nagaoka. 1984. "Effect of Cluster Thinning and Vineyard Yields on Grape and Wine Composition and Wine Quality of Cabernet Sauvignon." *American Journal of Enology and Viticulture* 35 (1):30–34.

Pacottet, P. 1905. *Viticulture.* Paris: J.-B. Baillière et fils.

Paguierre, M. 1828. *Classification and Description of the Wines of Bordeaux.* Edinburgh: W. Blackwood.

Palliotti, A. 2012. "A New Closing Y-shaped Training System for Grapevines." *Australian Journal of Grape and Wine Research* 18 (1):57–63. doi: 10.1111/j.1755–0238.2011.00171.x.

Palliotti, A., et al. 2011. "Early Leaf Removal to Improve Vineyard Efficiency: Gas Exchange, Source-to-Sink Balance, and Reserve Storage Responses." *American Journal of Enology and Viticulture* 62 (2):219–28. doi: 10.5344/ajev.2011.10094.

Pape, E. 2007. "French Wines Are Fighting Back." *Newsweek,* April 8, 2007. http://www.newsweek.com/french-wines-are-fighting-back-97655.

Parker, R. M. 1990. *Burgundy: A Comprehensive Guide to the Producers, Appellations, and Wines.* New York: Simon & Schuster.

———. 1997. *The Wines of the Rhône Valley.* New York: Simon & Schuster.

———. 2003. *Bordeaux: A Consumer's Guide to the World's Finest Wines.* New York: Simon & Schuster.

———. 2008. *Parker's Wine Buyer's Guide: The Complete, Easy-to-Use Reference on Recent Vintages, Prices, and Ratings for More Than 8,000 Wines from All the Major Wine Regions.* 7th ed. New York: Simon & Schuster.

———. 2010. *Burgundy: A Comprehensive Guide to the Producers, Appellations, and Wines.* New York: Simon & Schuster.

Parker, R. M., and P. A. Rovani. 2002. *Parker's Wine Buyer's Guide.* New York: Simon & Schuster.

Parkes, E. A., and C. Wollowicz. 1870. "Experiments on the Action of Red Bordeaux Wine (Claret) on the Human Body." *Proceedings of the Royal Society of London* 19 (123–29):73–89. doi: 10.1098/rspl.1870.0017.

Parr, W. V., et al. 2011. "Representation of Complexity in Wine: Influence of Expertise." *Food Quality and Preference* 22 (7):647–60.

Partridge, N. L. 1925. "The Fruiting Habits and Pruning of the Concord Grape." *Michigan State College of Agriculture and Applied Science, Agric Expt Stati Tech Bull Horticultural Section* 69:1–39.

Passmore, N. 2009. "Stressed-Out Vines Make Better Wines." *Bloomberg Businessweek,* April 2, 2009. http://www.businessweek.com/lifestyle/content/apr2009/bw2009042_512755.htm.

Pastor, M., et al. 2007. "Productivity of Olive Orchards in Response to Tree Density." *Journal of Horticultural Science & Biotechnology* 82 (4):555–62.

Paul, H. W. 1996. *Science, Vine, and Wine in Modern France.* Cambridge and New York: Cambridge University Press.

———. 2005. Review of *When Champagne Became French: Wine and the Making of a National Identity,* by Kolleen Guy. *Journal of Modern History* 77 (4):1104–8. doi: 10.1086/499865.

Pereira, G. E., et al. 2005. "H-1 NMR and Chemometrics to Characterize Mature Grape Berries in Four Wine-Growing Areas in Bordeaux, France." *Journal of Agricultural and Food Chemistry* 53 (16):6382–89. doi: 10.1021/Jf058058q.

Peterlunger, E., et al. 2002. "Effect of Training System on Pinot noir Grape and Wine Composition." *American Journal of Enology and Viticulture* 53 (1):14–18.

Petit, M. 1985. *Determinants of Agricultural Policies in the United States and the European Community.* Washington, DC: International Food Policy Research Institute.

Petrie, P. R., et al. 2000. "Growth and Dry Matter Partitioning of Pinot noir (Vitis vinifera L.) in Relation to Leaf Area and Crop Load." *Australian Journal of Grape and Wine Research* 6:40–45.

Peynaud, É. 1987. *The Taste of Wine: The Art Science of Wine Appreciation.* New York: Wiley.

Phillips, R. 2003. *Wine: Overview.* http://www.encyclopedia.com/topic/wine.aspx.

Phillips, R. L. 2014. "Green Revolution: Past, Present, and Future." In *Encyclopedia of Agriculture and Food Systems,* edited by Neal K. Van Alfen. San Diego: Elsevier.

Pierce, N. B. 1892. *The California Vine Disease: A Preliminary Report of Investigations.* Washington, DC: Government Printing Office.

Pinker, S. 1997. *How the Mind Works.* New York: W. W. Norton.

Pinney, T. 1989. *A History of Wine in America: From the Beginnings to Prohibition.* Berkeley: University of California Press.

———. 2005. *A History of Wine in America: From Prohibition to the Present.* Berkeley: University of California Press.

Pitte, J. R., and B. DeBevoise. 2008. *Bordeaux/Burgundy: A Vintage Rivalry.* Berkeley: University of California Press.

Plassmann, H., et al. 2008. "Marketing Actions Can Modulate Neural Representations of Experienced Pleasantness." *Proceedings of the National Academy of Sciences of the United States of America* 105 (3):1050–54.

Pliny. *Natural History.* Vol. 4. Translated by H. Rackham. Loeb Classical Library. Cambridge, MA: Harvard University Press, 1968.

———. *The Natural History of Pliny.* Translated by John Bostock and Henry Thomas Riley. Vol. 3. London: H. G. Bohn, 1855.

Pomerol, C. 1989. *The Wines and Winelands of France: Geological Journeys.* Paris: Éd du BRGM.

Pomerol, C., and France Bureau de recherches géologiques et minières. 1989. *The Wines and Winelands of France: Geological Journeys.* Orleans: Éditions du BRGM.

Poni, S., et al. 2007. "The Issue of Canopy Efficiency in the Grapevine: Assessment and Approaches for Its Improvement." *Acta Hort. (ISHS)* 754:163–74.

———. 2009. "Effects of Pre-bloom Leaf Removal on Growth of Berry Tissues and Must Composition in Two Red Vitis vinifera L. Cultivars." *Australian Journal of Grape and Wine Research* 15 (2):185–93. doi: 10.1111/j.1755-0238.2008 .00044.x.

Pool, R. M. 2004. "Vineyard Balance—What Is It? Can It Be Achieved?" *Acta Hort. (ISHS)* 640:285–302.

Popper, K. R. 1963. *Conjectures and Refutations: The Growth of Scientific Knowledge.* New York: Routledge.

Possner, D. R. E., and W. M. Kliewer. 1985. "The Localization of Acids, Sugars, Potassium, and Calcium in Developing Grape Berries." *Vitis* 24 (4): 229–40.

Poudel, P. R., et al. 2009. "Influence of Temperature on Berry Composition of Interspecific Hybrid Wine Grape 'Kadainou R-1' (Vitis ficifolia var. ganebu × V. vinifera 'Muscat of Alexandria')." *Journal of the Japanese Society for Horticultural Science* 78 (2):169–74.

Prajitna, A., et al. 2007. "Influence of Cluster Thinning on Phenolic Composition, Resveratrol, and Antioxidant Capacity in Chambourcin Wine." *American Journal of Enology and Viticulture* 58 (3):346–50.

Preszler, T., et al. 2013. "Cluster Thinning Reduces the Economic Sustainability of Riesling Production." *American Journal of Enology and Viticulture* 64 (3):333-41.

Prial, F. J. 1979. "Wine Talk." *New York Times,* March 7, 1979, 1-C10.

———. 1989. "Wine: The Battle of 1855." *New York Times,* August 20, 1989.

Priestley, J. 1775. *Experiments and Observations on Different Kinds of Air.* Vol. 2. London: J. Johnson.

———. 1777. *Experiments and Observations on Different Kinds of Air.* Vol. 3. London: J. Johnson.

Primack, R. B., and P. Hall. 1990. "Costs of Reproduction in the Pink Ladys-Slipper Orchid—a 4-Year Experimental-Study." *American Naturalist* 136 (5):638–56.

Prioreschi, P. 1998. *A History of Medicine: Roman Medicine.* Lewiston: Edwin Mellen Press.

Pruess, K. P. 1983. "Day-Degree Methods for Pest-Management." *Environmental Entomology* 12 (3):613–19.

Quandt, R. E. 2007a. "On Wine Bullshit: Some New Software?" *Journal of Wine Economics* 2 (2):129-35.

———. 2007b. Review of *The Accidental Connoisseur: An Irreverent Journey through the Wine World,* by Lawrence Osborne." *Journal of Wine Economics* 2 (2):222–23. doi:10.1017/S1931436100000481.

Ramsey, F. P. 1929. "General Propositions and Causality." In *F. P. Ramsey: The Foundations of Mathematics and Other Logical Essays*, edited by R. B. Braithwaite, 238. London: Routledge and Kegan Paul.

Ravaz, L. 1902. "Sur la cause de la brunissure." *Le Progrès Agricole et Viticole* 38:481–86.

———. 1903. "Sur la brunissure de la vigne." *C. R. Acad. Sci.* 136:1276–78.

———. 1904a. *La brunissure de la vigne: Cause, conséquences, traitement.* Montepellier: Coulet et Fils.

———. 1904b. "Recherches sur la brunissure de la vigne." *Journal d'Agriculture Pratique* 7:611–12.

———. 1906. "Influence de la surproduction sur la végétation de la vigne." *Ann. Ecole. Agric. Montpellier* 2 (6):5–41.

———. 1908. "Influence des opérations culturales sur la végétation et la production de la vigne." *Ann. de l'École Nat. d'Agric. de Montpellier* 8:231–92.

———. 1909. *Influence des opérations culturales sur la végétation et la production de la vigne.* Montpellier: Coulet et fils.

———. 1911. "L'effeuillage de la vigne." *Ann. l'École Nat. Agric. Montpellier* 11:216–44.

———. 1935. "Factors of Quality (in Wine) and Their Relation to Agricultural Practice." *Progrès Agricole et Viticole* 104:489–94.

Ravaz, L., et al. 1933. "Researches on rougeau of the Vine." *Annales Agronomiques* 3 (2):225–31.

Ravaz, L., and G. Verge. 1925. "On the Effect of Fertilizers on the Health of the Vine." *Ann. Ec. Agric.* 18 (4):237–44.

Réaumur, M. 1735. "Observations du thermometre, faites a Paris pendant l'année." *Mem. Acad. Roy. Sci. Paris,* 737–54.

Redding, C. 1851. *A History and Description of Modern Wines.* London: Whittaker, Treacher, & Arnot.

Rendig, V. V., and H. M. Taylor. 1989. *Principles of Soil-Plant Interrelationships.* New York: McGraw-Hill.

Reynolds, A. G. 1989. "Riesling Grapes Respond to Cluster Thinning and Shoot Density Manipulation." *Journal of the American Society for Horticultural Science* 114 (3):364–68.

Reynolds, A. G., et al. 1994a. "Fruit Environment and Crop Level Effects on Pinot-noir 1: Vine Performance and Fruit Composition in British-Columbia." *American Journal of Enology and Viticulture* 45 (4):452–59.

———. 1994b. "Shoot Density Affects Riesling Grapevines 2: Wine Composition and Sensory Response." *Journal of the American Society for Horticultural Science* 119 (5):881–92.

———. 1995. "Impact of Training System and Vine Spacing on Vine Performance and Berry Composition of Chancellor." *American Journal of Enology and Viticulture* 46 (1):88–97.

———. 1996a. "Fruit Environment and Crop Level Effects on Pinot noir 3: Composition and Descriptive Analysis of Oregon and British Columbia Wines." *American Journal of Enology and Viticulture* 47 (3):329–39.

———. 1996b. "Impact of Training System, Vine Spacing, and Basal Leaf Removal on Riesling, Vine Performance, Berry Composition, Canopy Microclimate, and

Vineyard Labor Requirements." *American Journal of Enology and Viticulture* 47 (1):63–76.

———. 2004a. "Evaluation of Winegrapes in British Columbia: New Cultivars and Selections from Germany and Hungary." *Horttechnology* 14 (3):420–36.

———. 2004b. "Impact of Training System and Vine Spacing on Vine Performance, Berry Composition, and Wine Sensory Attributes of Riesling." *American Journal of Enology and Viticulture* 55 (1):96–103.

———. 2004c. "Impact of Training System and Vine Spacing on Vine Performance, Berry Composition, and Wine Sensory Attributes of Seyval and Chancellor." *American Journal of Enology and Viticulture* 55 (1):84–95.

———. 2007. "Magnitude of Viticultural and Enological Effects II: Relative Impacts of Cluster Thinning and Yeast Strain on Composition and Sensory Attributes of Chardonnay Musque." *American Journal of Enology and Viticulture* 58 (1):25–41.

Reynolds, A.G., and J.E.V. Heuvel. 2009. "Influence of Grapevine Training Systems on Vine Growth and Fruit Composition: A Review." *American Journal of Enology and Viticulture* 60 (3):251–68.

Reynolds, A.G., and D.A. Wardle. 1994. "Impact of Training System and Vine Spacing on Vine Performance and Berry Composition of Seyval-Blanc." *American Journal of Enology and Viticulture* 45 (4):444–51.

Ribéreau-Gayon, P. 2000. "The Microbiology of Wine and Vinifications." In *Handbook of Enology*. Chichester, UK, and New York: Wiley.

Rizzini, F. M., et al. 2009. "Postharvest Water Loss Induces Marked Changes in Transcript Profiling in Skins of Wine Grape Berries." *Postharvest Biology and Technology* 52; 247–253.

Robertson, M., et al. 2009. "Tools for Managing Fruit Composition of Pinot Noir in Cool Climates." In *Final Report to the Grape and Wine Research and Development Corporation*. Tasmania: Tasmanian Institute of Agricultural Research, University of Tasmania, Australia.

Robinet, É. 1877. *Manuel général des vins: Fabrication des vins mousseux*. Paris: A. Lemoine.

Robinson, J. 1999. *The Oxford Companion to Wine*. Oxford: Oxford University Press.

———. 2002a. *Napa and That Conaway Book*. JancisRobinson.com, November 22, 2002. http://www.jancisrobinson.com/articles/jr7004.html.

———. 2002b. *Why I Love Burgundy*. JancisRobinson.com, February 2, 2002. http://www.jancisrobinson.com/articles/jr419.html.

———. 2007. *Irrigation Now Official in France*. JancisRobinson.com, April 11, 2007. http://www.jancisrobinson.com/articles/irrigation-now-official-in-france.

Robinson, J., and J. Baldwin. 2000. *How to Taste: A Guide to Enjoying Wine*. New York: Simon & Schuster.

Roby, G. 2001. "The Roles of Water Deficits and Berry Size in the Phenolic Composition of Cabernet Sauvignon Grapes and Wines." Master's thesis, Viticulture and Enology, University of California, Davis.

Roby, G., et al. 2004. "Berry Size and Vine Water Deficits as Factors in Winegrape Composition: Anthocyanins and Tannins." *Australian Journal of Grape and Wine Research* 10 (2):100–107.

Roby, G., and M. A. Matthews. 2004. "Relative Proportions of Seed, Skin, and Flesh in Ripe Berries from Cabernet Sauvignon Grapevines Grown in a Vineyard Either Well Irrigated or under Water Deficit." *Australian Journal of Grape and Wine Research* 10 (1):74–82.

Rodier, C. 1949. *Le Clos de Vougeot.* Marseille: Laffitte Reprints.

Romero-Cascales, I., et al. 2005. "Differences in Anthocyanin Extractability from Grapes to Wines According to Variety." *American Journal of Enology and Viticulture* 56 (3):212–19.

Rosenthal, J. 1990. "The Development of Irrigation in Provence, 1700–1860: The French Revolution and Economic Growth." *Journal of Economic History* 50 (3):615–38. doi: 10.1017/S0022050700037189.

Rosenthal, N. I. 2008. *Reflections of a Wine Merchant.* 1st ed. New York: Farrar, Straus and Giroux.

Ross, W. D., and J. A. Smith. 1912. *Works of Aristotle: De partibus animalium, Book III, translation by W. Ogle.* Oxford: Clarendon Press.

Roudié, P. 1988. *Vignobles et vignerons du Bordelais: 1850–1980.* Paris: Éditions du Centre national de la recherche scientifique.

———. 2001. "Vous avez dit terroir? Essai sur l'évolution d'un concept ambigu." *Journal International des Sciences de la Vigne et du Vin, n° hors série* 7.

Roux, S. 2009. *Paris in the Middle Ages.* The Middle Ages. Philadelphia: University of Pennsylvania Press.

Royal Society of Arts. 1764. *Museum Rusticum Et Commerciale: Or, Select Papers on Agriculture, Commerce, Arts, and Manufactures.* London: R. Davis.

Ryona, I., et al. 2008. "Effects of Cluster Light Exposure on 3-Isobutyl-2-Methoxypyrazine Accumulation and Degradation Patterns in Red Wine Grapes (Vitis vinifera L. Cv. Cabernet Franc)." *Journal of Agricultural and Food Chemistry* 56 (22):10838–46.

Sachs, J., et al. 1890. *History of Botany (1530–1860).* Oxford: Clarendon Press.

Sadras, V. O., et al. 2013. "Effects of Elevated Temperature in Grapevine I: Berry Sensory Traits." *Australian Journal of Grape and Wine Research* 19 (1):95–106.

Sadras, V. O., and M. A. Moran. 2012. "Elevated Temperature Decouples Anthocyanins and Sugars in Berries of Shiraz and Cabernet Franc." *Australian Journal of Grape and Wine Research* 18 (2):115–22.

Sadras, V. O., and P. R. Petrie. 2011. "Climate Shifts in South-Eastern Australia: Early Maturity of Chardonnay, Shiraz, and Cabernet Sauvignon Is Associated with Early Onset Rather Than Faster Ripening." *Australian Journal of Grape and Wine Research* 17 (2):199–205.

Santesteban, L. G., et al. 2010. "Vegetative Growth, Reproductive Development, and Vineyard Balance." In *Methodologies and Results in Grapevine Research,* edited by S. Delrot et al. Dordrecht: Springer.

Sapolsky, R. M. 2012. "Aspiration Makes Us Human." *Scientific American,* September 2012.

Schneider, P. E. 1957. "France's Wine Casks Are in the Red." *New York Times,* December 8, 1957.

Schultz, H. R. 1993. "Photosynthesis of Sun and Shade Leaves of Field-Grown Grapevine (Vitis-Vinifera L) in Relation to Leaf Age—Suitability of the Plastochron Concept for the Expression of Physiological Age." *Vitis* 32 (4): 197–205.

———. 2000. "Climate Change and Viticulture: A European Perspective on Climatology, Carbon Dioxide, and UV-B Effects." *Australian Journal of Grape and Wine Research* 6:2–12.

———. 2003. "Extension of a Farquhar Model for Limitations of Leaf Photosynthesis Induced by Light Environment, Phenology, and Leaf Age in Grapevines (Vitis vinifera L. cvv. White Riesling and Zinfandel)." *Functional Plant Biology* 30 (6):673–87.

Schultz, H. R., et al. 1996. "Photosynthetic Duration, Carboxylation Efficiency, and Stomatal Limitation of Sun and Shade Leaves of Different Ages in Field-Grown Grapevine (Vitis vinifera L)." *Vitis* 35 (4):169–76.

Schultz, H. R., and M. A. Matthews. 1988. "Vegetative Growth Distribution during Water Deficits in Vitis-Vinifera L." *Australian Journal of Plant Physiology* 15 (5):641–56.

———. 1993. "Xylem Development and Hydraulic Conductance in Sun and Shade Shoots of Grapevine (Vitis-Vinifera L)—Evidence That Low-Light Uncouples Water Transport Capacity from Leaf-Area." *Planta* 190 (3): 393–406.

Schuttler, P. A. 2013. "Influencing Factors on Aromatic Typicality of Wines from Vitis vinifera L. cv. Riesling—Sensory, Chemical, and Viticultural Insights." PhD diss., Graduate School of Life Sciences and Health, University Bordeaux Segalen and University of Giessen.

Scienza, A., et al. 1978. "Relationships between Seed Number, Gibberellin and Abscisic-Acid Levels, and Ripening in Cabernet Sauvignon Grape Berries." *Vitis* 17 (4):361–68.

———. 1996. "A Multi-disciplinary Study of the Vineyard Ecosystem to Optimize Wine Quality." *Acta Hort. (ISHS)* 427:347–62.

Scott-Moncrieff, C. C. 1868. *Irrigation in Southern Europe: Being the Report of a Tour of Inspection of the Irrigation Works of France, Spain, and Italy, Undertaken in 1867–68 for the Government of India.* London: Spon.

Seguin, G. 1975. "Alimentation en eau de la vigne et composition chimique des mouts dans les Grands Crus du Medoc: Phenomenes de regulation." *Connaissance de la Vigne et du Vin* 9:23–34.

———. 1986. "'Terroirs' and Pedology of Vinegrowing." *Experientia* 42: 861–73.

Seneca, Lucius Annaeus. *Letters from a Stoic.* Translated by R. Campbell. London: Penguin, 2004.

Shanken, M. R., and T. Matthews. 2010. "Who Is America's Greatest Winemaker?" *Wine Spectator,* July 31, 2010.

Shaulis, N. J. 1982. "Responses of Grapevines and Grapes to Spacing of and within Canopies." *Proceedings, Grape and Wine Centennial Symposium, June 1980, University of California, Davis,* 353–61.

Shaulis, N., and K. Kimball. 1955. "Effect of Plant Spacing on Growth and Yield of Concord Grapes." *Proceedings of the American Society for Horticultural Science* (66):192–200.

Shaw, T. G. 1864. *Wine, the Vine, and the Cellar.* London: Longman, Green, Longman, Roberts, & Green.

Shellie, K. C. 2007. "Viticultural Performance of Red and White Wine Grape Cultivars in Southwestern Idaho." *Horttechnology* 17 (4):595-603.

Shermer, M. 2008. "Adam's Maxim and Spinoza's Conjecture." *Scientific American* 298 (3):36–37.

Simon, J. 2003. *Discovering Wine.* New York: Simon and Schuster.

Simpson, J. 2005. "The Midi, Bordeaux, and Champagne: Cooperation and Conflicts; Institutional Innovation in France's Wine Markets, 1870–1911." *Business History Review* 79 (3):527–58.

Singleton, V. L. 1966. "The Total Phenolic Content of Grape Berries during the Maturation of Several Varieties." *American Journal of Enology and Viticulture* 17 (2):126–34.

———. 1972. "Effects on Red Wine Quality of Removing Juice before Fermentation to Simulate Variation in Berry Size." *American Journal of Enology and Viticulture* 23 (3):106–13.

Singleton, V. L., and J. P. Esau. 1969. *Phenolic Substances in Grapes and Wine and Their Significance.* Advances in Food Research. New York: Academic Press.

Sinton, T. H., et al. 1978. "Grape Juice Indicators for Prediction of Potential Wine Quality I: Relationship between Crop Level, Juice, and Wine Composition, and Wine Sensory Ratings and Scores." *American Journal of Enology and Viticulture* 29 (4):267–71.

Skinner, J. S. 1826. *American Farmer* 8 (48):381.

Smart, D. R., et al. 2006. "Physiological Changes in Plant Hydraulics Induced by Partial Root Removal of Irrigated Grapevine (Vitis vinifera cv. Syrah)." *American Journal of Enology and Viticulture* 57 (2):201–9.

Smart, R. E. 1985. "Principles of Grapevine Canopy Microclimate Manipulation with Implications for Yield and Quality—a Review." *American Journal of Enology and Viticulture* 36 (3):230–39.

———. 1992. "Canopy Management." In *Viticulture,* vol. 2, *Practices,* edited by B. G. Coombe and P. R. Dry, 85–103. Adelaide: Winetitles.

Smart, R. E., et al. 1990. "Canopy Management to Improve Grape Yield and Wine Quality—Principles and Practices." *South African Journal of Enology and Viticulture* 11:3–17.

Smart, R. E., and B. G. Coombe. 1983. "Water Relations of Grapevines." In *Additional Woody Crop Plants,* edited by T. T. Kozlowski. New York: Elsevier Science.

Smith, F. 2004. "Do Greater Yields Mean Lesser Wines? (Wine Quality)." *Wines and Vines* 85 (9):48-51.

Soubeyrand, E., et al. 2014. "Nitrogen Supply Affects Anthocyanin Biosynthetic and Regulatory Genes in Grapevine cv. Cabernet-Sauvignon Berries." *Phytochemistry* 103:38–49.

Spayd, S. E., et al. 1993. "Nitrogen-Fertilization of White-Riesling in Washington—Effects on Petiole Nutrient Concentration, Yield, Yield Components, and Vegetative Growth." *American Journal of Enology and Viticulture* 44 (4):378–86.

Sprat, T. 1667. *The history of the Royal-society of London.* London: Printed by T. R. for J. Martyn etc.

Stefanini, M., et al. 2000. "Adaptation of Some Cabernet-Sauvignon Clones to the Environmental Conditions of North-Eastern Italian Growing Areas." *Acta Hort. (ISHS)* 528:779–84.

Stein, R. I., and C. J. Nemeroff. 1995. "Moral Overtones of Food: Judgments of Others Based on What They Eat." *Personality and Social Psychology Bulletin* 21 (5):480–90.

Steiner, R. 1958. *The Agriculture Course.* London: Bio-Dynamic Agricultural Association, Rudolf Steiner House.

Stevenson, T. 2007. *Champagne's 6 Billion Expansion.* http://www.wine-pages.com/guests/tom/champagne-expands.htm.

Storey, R. 1987. "Potassium Localization in the Grape Berry Pericarp by Energy-Dispersive X-Ray-Microanalysis." *American Journal of Enology and Viticulture* 38 (4):301–9.

Stroup, A. 1990. *A Company of Scientists: Botany, Patronage, and Community at the Seventeenth-Century Parisian Royal Academy of Sciences.* Berkeley: University of California Press.

Suklje, K., et al. 2012. "Classification of Grape Berries According to Diameter and Total Soluble Solids to Study the Effect of Light and Temperature on Methoxypyrazine, Glutathione, and Hydroxycinnamate Evolution during Ripening of Sauvignon blanc (Vitis vinifera L.)." *Journal of Agricultural and Food Chemistry* 60 (37):9454–61.

Sumby, K. M., et al. 2010. "Microbial Modulation of Aromatic Esters in Wine: Current Knowledge and Future Prospects." *Food Chemistry* 121 (1):1–16. doi: 10.1016/j.foodchem.2009.12.004.

Sun, B. S., et al. 1999. "Transfer of Catechins and Proanthocyanidins from Solid Parts of the Grape Cluster into Wine." *American Journal of Enology and Viticulture* 50 (2):179–84.

Swinchatt, J. P., and D. G. Howell. 2004. *The Winemaker's Dance: Exploring Terroir in the Napa Valley.* Berkeley: University of California Press.

Tanzer, S. 2006. *The Wine Access Buyer's Guide: The World's Best Wines & Where to Find Them.* New York: Sterling Publishing.

Tassie, E., and B. M. Freeman. 1992. "Pruning." In *Viticulture,* vol. 2, *Practices,* edited by B. G. Coombe and P. R. Dry. Adelaide: Winetitles.

Terrier, A., et al. 1972. "Teneurs en composes terpeniques des raisins de Vitis vinifera." *C.R. Acad. Sci. Paris Ser. D* 275 (8):941–44.

Tesic, D., et al. 2007. "Influence of Vineyard Floor Management Practices on Grapevine Vegetative Growth, Yield, and Fruit Composition." *American Journal of Enology and Viticulture* 58 (1):1–11.

Thomas, J. E., and C. Barnard. 1937. "Fruit Bud Studies III: The Sultana; Some Relations between Shoot Growth, Chemical Composition, Fruit Bud Formation, and Yield." *Journal of the Council for Scientific and Industrial Research Australia* 10:143–57.

Thudichum, J. L. W., and August Dupré. 1872. *A Treatise on the Origin, Nature, and Varieties of Wine; Being a Complete Manual of Viticulture and Oenology.* London and New York: Macmillan.

Tobin, S. 1996. *The Cistercians: Monks and Monasteries of Europe.* Woodstock: Overlook Press.

Tomasi, D., et al. 2014. *The Power of the Terroir: The Case Study of Prosecco Wine.* New York: Springer Basel.

Tomasik, T. J. 2001. "Certeau à-la-Carte: Translating Discursive Terroir in the 'Practice of Everyday Life, Living and Cooking.'" *South Atlantic Quarterly* 100 (2):519–42. doi: 10.1215/00382876-100-2-519.

Tonietto, J., and A. Carbonneau. 2004. "A Multicriteria Climatic Classification System for Grape-Growing Regions Worldwide." *Agricultural and Forest Meteorology* 124 (1–2):81–97.

Tournier, A. 1907. "The Failure of Vines." *Agricultural Journal of the Cape of Good Hope* 30:95–100.

Toussaint-Samat, M., and A. Bell. 1992. *A History of Food.* Cambridge: Blackwell Reference.

Trought, M. C. T., and S. J. Tannock. 1996. "Berry Size and Soluble Solids Variation within a Bunch of Grapes." *Proceedings of the 4th International Symposium on Cool Climate Viticulture and Enology, Rochester, New York,* V70–V73.

Tull, J. 1762. *Horse-hoeing husbandry.* London: Printed for A. Millar.

Turkington, C. R., et al. 1980. "A Spacing, Trellising, and Pruning Experiment with Muscat Gordo Blanco Grapevines." *American Journal of Enology and Viticulture* 31 (3):298–302.

Ulin, R. C. 1987. "Writing and Power—the Recovery of Winegrowing Histories in the Southwest of France." *Anthropological Quarterly* 60 (2):77–82. doi: 10.2307/3317998.

———. 1995. "Invention and Representation as Cultural Capital—Southwest French Winegrowing History." *American Anthropologist* 97 (3):519–27. doi: 10.1525/aa.1995.97.3.02a00100.

Unwin, T. 1996. *Wine and the Vine: An Historical Geography of Viticulture and the Wine Trade.* New York: Taylor & Francis.

———. 2000. "The Viticultural Geography of France in the 17th Century According to John Locke/La géographie viticole de la France au XVIIe siècle selon John Locke." *Annales de Géographie,* 395–414.

Vaadia, Y., and A. N. Kasimatis. 1961. "Vineyard Irrigation Trials." *American Journal of Enology and Viticulture* 12:88–98.

Vance, A. J., and P. Skinkis. 2013. *Understanding Vine Balance: An Important Concept in Vineyard Management.* Corvallis: Oregon State University Extension Service.

van Leeuwen, C., et al. 2004. "Influence of Climate, Soil, and Cultivar on Terroir." *American Journal of Enology and Viticulture* 55 (3):207–17.

Vasconcelos, M. C., and S. Castagnoli. 2000. "Leaf Canopy Structure and Vine Performance." *American Journal of Enology and Viticulture* 51 (4):390–96.

Vaudour, E. 2001. *Les terroirs viticoles.* Paris: Dunod.

Viala, P., and L. Ravaz. 1888. *Le black rot et le Coniothyrium diplodiella.* Paris: A. Delahaye et E. Lecrosnier.

Vidal, S., et al. 2004. "The Mouth-Feel Properties of Polysaccharides and Anthocyanins in a Wine-Like Medium." *Food Chemistry* 85 (4):519–25.

Vincent, C. 1882. *Report of a Tour of Inspection of Irrigation Works in Southern France and Italy: And of Some of the Principal Masonry Dams in France.* Madras: E. Keys at the Government Press.

Virgil. *The Georgics.* Translated by A.S. Kline. Book 2, *Arboriculture and Viniculture.* http://www.poetryintranslation.com/PITBR/Latin/VirgilGeorgicsII .htm#_Toc533843188.

Walker, R.R., et al. 2005. "Shiraz Berry Size in Relation to Seed Number and Implications for Juice and Wine Composition." *Australian Journal of Grape and Wine Research* 11 (1):2–8. doi: 10.1111/j.1755-0238.2005.tb00273.x.

Wang, J.Y. 1960. "A Critique of the Heat Unit Approach to Plant-Response Studies." *Ecology* 41 (4):785–90.

Watada, A.E., et al. 1984. "Terminology for the Description of Developmental Stages of Horticultural Crops." *Hortscience* 19 (1):20–21.

Watson, B.T., et al. 1988. "Evaluation of Pinot noir clones in Oregon." *Proceedings of the Second International Symposium for Cool Climate Viticulture and Oenology,* edited by R.E. Smart, R.J. Thorton, S.B. Rodriquez, and J.E. Young, 276–78.

Watson, J.M., and K. Riha. 2011. "Telomeres, Aging, and Plants: From Weeds to Methuselah—a Mini-Review." *Gerontology* 57 (2):129–36. doi: 10.1159/000310174.

Weaver, R.J. 1973. "Effect of Chlormequat [(2-Chloroethyl)-Trimethylammonium Chloride] on Small-Berried Wine Grapes." *American Journal of Enology and Viticulture* 24 (2):69–71.

Weaver, R.J., et al. 1957. "Preliminary Report on Effect of Level of Crop on Development of Color in Certain Red Wine Grapes." *American Journal of Enology and Viticulture* 8 (4):157–66.

———. 1961. "Effect of Level of Crop on Vine Behavior and Wine Composition in Carignane and Grenache Grapes." *American Journal of Enology and Viticulture* 12 (4):175–84.

Weaver, R.J., and R.M. Pool. 1968. "Effect of Various Levels of Cropping on Vitis Vinifera Grapevines." *American Journal of Enology and Viticulture* 19 (3):185–93.

Webb, R.A., and D.G. Hallas. 1966. "Effect of Iron Supply on Strawberry Var Royal Sovereign." *Journal of Horticultural Science & Biotechnology* 41 (2):179.

Webster, D.R., et al. 1993. "Influence of Vineyard Nitrogen-Fertilization on the Concentrations of Monoterpenes, Higher Alcohols, and Esters in Aged Riesling Wines." *American Journal of Enology and Viticulture* 44 (3):275–84.

Weigend, G.G. 1954. "The Basis and Significance of Viticulture in Southwest France." *Annals of the Association of American Geographers* 44 (1):75–101.

Wenke, K., et al. 2010. "Belowground Volatiles Facilitate Interactions between Plant Roots and Soil Organisms." *Planta* 231 (3):499–506. doi: 10.1007/ s00425-009-1076-2.

West, J.W. 2003. "Effects of Heat-Stress on Production in Dairy Cattle." *Journal of Dairy Science* 86 (6):2131–44.

Wetmore, C.A. 1884. "Second Annual Report of the Chief Executive Viticultural Officer to the Board of State Viticultural Commissioners, for the Year

1882–3 and 1883–4." In *Journals of the Legislature of the State of California,* vol. 5. Sacramento: State Office.

Weyand, K.M., and H.R. Schultz. 2006a. "Long-Term Dynamics of Nitrogen and Carbohydrate Reserves in Woody Parts of Minimally and Severely Pruned Riesling Vines in a Cool Climate." *American Journal of Enology and Viticulture* 57 (2):172–82.

———. 2006b. "Regulating Yield and Wine Quality of Minimal Pruning Systems through the Application of Gibberellic Acid." *Journal International des Sciences de la Vigne et du Vin* 40 (3):151–63.

Whalen, P. 2007a. "Burgundian Regionalism and French Republican: Commercial Culture at the 1937 Paris International. Exposition." *Cultural Analysis* (6):31–69.

———. 2007b. "'A Merciless Source of Happy Memories': Gaston Roupnel and the Folklore of Burgundian Terroir." *Journal of Folklore Research* 44 (1):21–40. doi: 10.2979/Jfr.2007.44.1.21.

———. 2009. "'Insofar as the Ruby Wine Seduces Them': Cultural Strategies for Selling Wine in Inter-war Burgundy." *Contemporary European History* 18 (1):67–98. doi: 10.1017/S0960777308004839.

———. 2010. "Jean-Robert Pitte, Bordeaux/Burgundy: A Vintage Rivalry." *H-France Review* 10 (65):294–302.

White, P.J., et al. 2009. "Relationships between Yield and Mineral Concentrations in Potato Tubers." *Hortscience* 44 (1):6–11.

White, R.D., et al. 2010. "The Dogmas of Nutrition and Cancer: Time for a Second (and Maybe Third) Look." *Annals of the New York Academy of Sciences* 1190:118–25. doi: 10.1111/j.1749–6632.2009.05271.x.

White, R.E. 2003. *Soils for Fine Wines.* New York: Oxford University Press.

———. 2009. *Understanding Vineyard Soils.* Oxford and New York: Oxford University Press.

Wiebe, J., and O.A. Bradt. 1973. "Fruit Yields and Quality in the Early Years of a Grape-Spacing Trial." *Canadian Journal of Plant Science* 53 (1):1 53–56.

Wildman, W.E., et al. 1976. "Improving Grape Yield and Quality with Depth-Controlled Irrigation." *American Journal of Enology and Viticulture* 27 (4):168–75.

Wilhelm, A.F. 1950. "Contributions to the Knowledge of Potassium Deficiency Symptoms in the Vine, Vitis vinifera L." *Phytopathologische Zeitschrift* 17 (3):240–65.

Wilkinson, L.P. 1978. *The Georgics of Virgil: A Critical Survey.* Cambridge: Cambridge University Press.

Williams, D., and R. Arnold. 1999. "Evaluation of Cabernet Sauvignon: Three Vine Spacings, Two Trellis Systems; Oakville District, Napa Valley." *Practical Winery & Vineyard Journal,* September/October 1999. http://www.practicalwinery.com/septoct99/mondavi.htm.

Williams, L.E., et al. 2010a. "The Effects of Applied Water at Various Fractions of Measured Evapotranspiration on Reproductive Growth and Water Productivity of Thompson Seedless Grapevines." *Irrigation Science* 28 (3): 233–43.

———. 2010b. "The Effects of Applied Water at Various Fractions of Measured Evapotranspiration on Water Relations and Vegetative Growth of Thompson Seedless Grapevines." *Irrigation Science* 28 (3):221–32.

Wilson, J. E. 1998. *Terroir: The Role of Geology, Climate, and Culture in the Making of French Wines.* Berkeley: University of California Press.

Wilson, M. L. 2011. "Expanding the Monopoly of Champagne, France." Paper presented at the 6th Annual Conference in World History and Economics, April 30, 2011, Appalachian State University, Boone, NC.

Winiarski, Warren. "Zut alors! The French like California Wine." *Wines & Vines,* April 1991. 72, 28.

Winkler, A. J. 1926. "Some Responses of Vitis vinifera to Pruning." *Hilgardia* 1:525–43.

———. 1931. "Pruning and Thinning Experiments with Grapes." *California Agricultural Experiment Station Bulletin* 519:1–56.

———. 1934. "Pruning Vinifera Grapevines." *California Agricultural Extension Service Circular* 89.

———. 1945. *Pruning Vinifera Grapevines.* Berkeley: University of California, College of Agriculture.

———. 1954. "Effects of Overcropping." *American Journal of Enology and Viticulture* 5:4–12.

———. 1958. "The Relation of Leaf Area and Climate to Vine Performance and Grape Quality." *American Journal of Enology* 9:10–23.

———. 1959. "The Effect of Vine Spacing at Oakville on Yields, Fruit Composition, and Wine Quality." *American Journal of Enology and Viticulture* 10 (1):39–43.

———. 1962. *General Viticulture.* Berkeley: University of California Press.

———. 1969. "Effect of Vine Spacing in an Unirrigated Vineyard on Vine Physiology Production and Wine Quality." *American Journal of Enology and Viticulture* 20 (1):7-15.

———. 1970. *General Viticulture.* Berkeley: University of California Press.

———. 1971. "Growth and Fruiting of Mission Grapevine under Conditions of Non-Competition—The Winkler Vine during Its First 10 Years." *American Journal of Enology and Viticulture* 22 (4):231-33.

Winkler, A. J., et al. 1974. *General Viticulture.* 2nd rev. ed. Berkeley: University of California Press.

Winkler, A. J., and M. A. Amerine. 1943. *Grape Varieties for Wine Production.* Berkeley: University of California Printing Office.

Wolf, T. K. 2011. "A Response from Tony Wolf to the Vine Spacing Question." *Grape Press* 27 (1): 7. http://www.virginiavineyardsassociation.com/wp-content/uploads/2011/06/GrapePress-April-2011.pdf.

Wolf, T. K., et al. 2003. "Response of Shiraz Grapevines to Five Different Training Systems in the Barossa Valley, Australia." *Australian Journal of Grape and Wine Research* 9 (2):82–95. doi: 10.1111/j.1755-0238.2003.tb00257.x.

Woods, A. F. 1899. "Brunissure of the Vine and Other Plants." *Science* 9 (223):508–10. doi: 10.2307/1626088.

Woods, T. E. 2005. *How the Catholic Church Built Western Civilization.* Washington, DC, and Lanham, MD: Regnery Pub.

Zacharewicz, E. 1934. "Combined Treatment against Vine 'Brunissure' and Chlorosis." *Progrès Agricole et Viticole* 102 (43):423–24.

Zacharkiw, B. 2010. "New Zealand's Pinot noirs Gaining a Solid Reputation." http://www.centralotagopinot.co.nz/news/13_New%20Zealand%E2%80%99s%20pinot%20noirs%20gaining%20a%20solid%20reputation.html.

Zamboni, M., et al. 1996. "Influence of Bud Number on Growth, Yield, Grape, and Wine Quality of 'Pinot gris', 'Pinot noir', and 'Sauvignon' (Vitis vinifera L.)." *Acta Hort. (ISHS)* 427:411–20.

Zhuang, S. J., et al. 2014. "Impact of Cluster Thinning and Basal Leaf Removal on Fruit Quality of Cabernet Franc (Vitis vinifera L.) Grapevines Grown in Cool Climate Conditions." *Hortscience* 49 (6):750–56.

Zimmerman, A. 2009. *Tasting Dirt: The Continuity of Terroir, First-Hand.* Culinate, September 21, 2009. http://www.culinate.com/articles/features/laura_parker_taste_of_place.

Zirkle, C. 1969. "Plant Hybridization and Plant Breeding in Eighteenth-Century American Agriculture." *Agricultural History* 43 (1):25–38. doi: 10.2307/4617624.

Zorrilla-Fontanesi, Y., et al. 2011. "Quantitative Trait Loci and Underlying Candidate Genes Controlling Agronomical and Fruit Quality Traits in Octoploid Strawberry (Fragaria × ananassa)." *Theoretical and Applied Genetics* 123 (5):755–78. doi: 10.1007/s00122-011-1624-6.

Index

Note: Page numbers in italic type refer to graphic material.

Milton Keynes UK
Ingram Content Group UK Ltd.
UKHW042051080924
447994UK00002BA/19/J